DISCARDED

THE DIVERSITY, COMPLEXITY, AND EVOLUTION OF HIGH TECH CAPITALISM

Distributors for North America:
Kluwer Academic Publishers
101 Philip Drive
Assinippi Park
Norwell, Massachusetts 02061 USA

Distributors for all other countries:
Kluwer Academic Publishers Group
Distribution Centre
Post Office Box 322
3300 AH Dordrecht, THE NETHERLANDS

Library of Congress Cataloging-in-Publication Data

A C.P.I. Catalogue record for this book is available from
the Library of Congress.

Copyright © 1995 by Kluwer Academic Publishers

All rights reserved. No part of this publication may be reproduced, stored in a retrieval system or transmitted in any form or by any means, mechanical, photo-copying, recording, or otherwise, without the prior written permission of the publisher, Kluwer Academic Publishers, 101 Philip Drive, Assinippi Park, Norwell, Massachusetts 02061.

Printed on acid-free paper.

Printed in the United States of America

THE DIVERSITY, COMPLEXITY, AND EVOLUTION OF HIGH TECH CAPITALISM

by

Sten Thore

IC² Institute
The University of Texas at Austin

Kluwer Academic Publishers
Boston/Dordrecht/London

For Margrethe

CONTENTS

FOREWORD	xi
CHAPTER 1 -- THE CHANGING FACE OF CAPITALISM	1
Stumbling Giants and a New Breed of Startup Companies	2
Economic Diversity	4
What is High Technology?	7
Technological Evolution at the Edge of Chaos	8
The Collision with the Environment	11
Where Did Marx Go Wrong?	13
The Freedom to Create	15
Bibliographic Notes	16
References	17
CHAPTER 2 -- NEW WEBS AND NEW CONSTITUENCIES	19
What is a Strategic Alliance?	20
Universities and Federal Laboratories: The New Breeding Grounds for High Tech Companies	22
References	25
CHAPTER 3 -- MESSAGES, IMAGES, AND ROBOTS	27
The Bewildering Future of the Telephone	30
Supercomputers and Parallel Processing	31
Artificial Intelligence	33
Robots: Slaves or Masters?	37
Bibliographic Notes	41
References	42
CHAPTER 4 -- THE TECHNOLOGY FRONTIER	45
The Rise and The Fall of the Passenger Train	45
What is Productivity?	49
Pushing the Technology Envelope	52
Complexity and Evolution: The Santa Fe Institute	56
Bibliographic Notes	59
References	59
CHAPTER 5 -- ECONOMIC LOGISTICS	61
The Dimensions and Mathematical Techniques of Economic Logistics	62
Innovation and Technological Change in an Industry Network	66
Budding and Cross-Fertilization	67
Bibliographic Notes	72
References	73

CHAPTER 6 -- OF PRINCES, DIPLOMACY, AND BATTLE CRIES ... 75
   The Strategies of Takeovers and Acquisitions ... 76
   Capital Gains: How to Make Them ... 83
   The Anatomy of a Hostile Takeover: RJR Nabisco ... 87
   Are Corporate Raiders Good or Bad for Industry? ... 89
   Bibliographic Notes ... 91
   References ... 91

CHAPTER 7 -- THE LONG WAVES: EVOLUTION OR CHAOS? ... 93
   The Long Swings ... 93
   Reversible and Irreversible Time ... 96
   The Ebbs and Tides of Basic Innovations ... 97
   The Diffusion of New Technology and New Products ... 99
   The Possibility of Chaos ... 101
   Turbulence and Shake-Out in U.S. Industry ... 105
   Bibliographic Notes ... 107
   References ... 107

CHAPTER 8 -- GENETIC SELECTION AND BIOTECHNOLOGY ... 109
   Early Technology on the Farm ... 109
   The Creation of New Varieties and New Species through Hybridization and Selection ... 110
   Pesticides and Herbicides ... 112
   Genetic Engineering on the Farm ... 116
   Genetically Engineered Drugs ... 118
   The Biotech Boom, and the Food and Drug Administration ... 120
   Ethical Questions: Are We Playing God? ... 121
   Bibliographic Notes ... 123
   References ... 124

CHAPTER 9 -- COMBUSTION, FISSION, AND FUSION ... 125
   Are We Running out of Oil? ... 125
   The Hidden Costs of Oil ... 127
   Coal and Natural Gas ... 130
   Don't Drink the Water and Don't Breathe the Air ... 132
   Nuclear Fission and Nuclear Fusion ... 136
   Yin and Yang ... 138
   Bibliographic Notes ... 139
   References ... 140

CHAPTER 10 -- VENTURES AND START-UP COMPANIES ... 143
   Silicon Valley: the First Technopolis ... 143
   The R&D Consortium ... 145
   The Business of Incubation: Breeding and Nurturing New Companies ... 149
   Venture Capital ... 150

Enterprise Zones	153
References	153

CHAPTER 11 -- RICHES FROM JUNK ... 155
   Background: The Market for Government Treasury Bonds ... 155
   Corporate Bonds: Investment Grade and Below Investment Grade ... 156
   Michael Milken: Financial Genius or a Criminal? ... 158
   The Importance of High-Risk Debt ... 159
   The Future of Junk ... 162
   Bibliographic Notes ... 163
   References ... 163

CHAPTER 12 -- THRIFTS AND BANKS IN TURMOIL ... 165
   The Savings and Loan Debacle ... 167
   Financial Innovation: Packaging and Re-packaging Mortgage Securities ... 169
   Electronic Money ... 170
   The Sorry Story of U.S. Banking ... 172
   Bibliographic Notes ... 176
   References ... 177

CHAPTER 13 -- THE SOCIAL MANAGEMENT OF TECHNOLOGY: WHERE ARE WE HEADED? ... 179
   The Blessings and Curses of Technology ... 179
   How Can Society Manage Technology ... 182
   Regulation ... 184
   Hearings and Licensing ... 185
   Industrial Policy ... 186
   Technological Infrastructure ... 188
   Commercialization Policy ... 189
   The Swarming of High Technology ... 190
   The Management of the Future ... 193
   Bibliographic Notes ... 194
   References ... 194

REFERENCES ... 197

INDEX OF CORPORATIONS AND EXECUTIVE OFFICERS ... 199

ABOUT THE AUTHOR ... 203

List of Text Boxes

THE VISION OF GEORGE KOZMETSKY	23
MILESTONES OF THE TELEPHONE	28
MILESTONES IN PHOTOGRAPHY	29
MILESTONES IN TELEVISION	33
A VISIT TO DELBERT TESAR'S LABORATORY	39
A. CHARNES AND W.W. COOPER	53
MAJOR CORPORATE TAKEOVERS IN THE U.S. SINCE 1985	80
ILYA PRIGOGINE -- SCIENTIST AND HUMANIST	104
A VISIT TO MALCOLM BROWN'S LABORATORY.	114
MILESTONES IN BIOTECHNOLOGY	119
A CHECKLIST OF ALTERNATIVE ENERGY SOURCES (ALTERNATIVE TO FOSSIL FUELS)	135
MILESTONES IN MODERN BANKING	166

Foreword

In his book "Jurassic Park" (and in the movie based on the book), Michael Crichton describes a crazed professor who through techniques of genetic engineering manages to recreate the dinosaurs and giant ferns of 65 million years past. Once the giant Tyrannosaurus Rex is brought to life, a powerful dynamics sets in: evolution. The prehistoric world embarks on a collision course with man.

Researching his book, Crichton had been reading up on paleontology and on the mathematical theory of evolution, catastrophes, and chaos. Crichton explains some of the twists of nonlinear mathematics that are rewriting not only thermodynamics, physics, and chemistry (that all grapple with evolving and turbulent processes) but also paleontology, genetics, medicine and even anthropology.

Collapse and chaos is not limited to prehistoric animal kingdoms and ancient civilizations. The collapse of the Soviet Union and the political and economic chaos in its aftermath demonstrate that modern civilizations are just as vulnerable.

This book aims at reexamining some main portions of the discipline of economics from the point of view of economic change and creativity. There are two aspects to this perspective. First, diversity and complexity. The range of different kinds of high technology products available to consumers and producers increases rapidly. Each product is the result of a long and complex production hierarchy. As these hierarchies grow, they deliver ever more diversified and complex high tech goods. Other hierarchies fall by the wayside. What will the new technologies of multimedia and interactive television look like? We do not know yet. Powerful new corporations will emerge. Others will crumble.

Second, self-organization and evolution. Living organisms spontaneously establish patterns and order. So does the economy. Some of the patterns of the economic world used to be large vertically integrated corporations and multinational corporations -- the corporate dinosaurs of the 1960s and 1970s. This structure is now falling apart (witness the plight of General Motors, IBM, and even of Sears and Roebuck). New industrial patterns are emerging: small narrowly specialized corporations, like Symbol Technologies making the fonts for computer printers, or IG Labs making genetic test kits. It is in these startups that the jobs of the future will be created.

I shall argue that the U.S. economy is currently experiencing a turbulent "phase transition", moving from one phase of capitalism to the next. New species of corporations and markets, and economic institutions are emerging. Out of the present chaos, a new and higher order of capitalism is crystallizing.

What will the future bring? As the many-splendored engine of capitalism gets into gear, there is no limit to the technological diversity and beauty that the economic goods of the future will bring. The instrument of change is technological creativity, the inventive human mind calling on the God-given creative powers deposited deep within her. But only those commercial products that fit into the self-organizing pattern of capitalism will succeed. Those that do not belong will fail.

My sincere thanks are due to Dr. Karl-Olof Faxén, economic adviser and former chief economist, Swedish Employers' Confederation, to Dr. Kenneth C. Land, chairman of the department of sociology, Duke University, and to Dr. Alfred L. Norman, professor in economics at the University of Texas at Austin. They kindly read my manuscript, criticized it, and gave me valuable advice. -- Technological developments and economic events in the U.S. as related in this book have been brought up to date as of early spring, 1995.

S. Th.

CHAPTER 1

The Changing Face of Capitalism

The last quarter of this century has witnessed the emergence of a new economic order. That new order is the capitalism of Intel Corp., Microsoft Corp., and Genentech Inc. It is the capitalism of a new kind of turbulent American entrepreneurship. In retrospect, in the course of world history, the emergence of this new economic order may very well turn out to be as important as, say, the advent of enlightenment or the renaissance. It marks the arrival of a new age -- an age of artificial intelligence, robotics, space research, biotech, and environmental technology.

The new capitalism is also the capitalism of the economic power-houses of the Far East: the Japanese superstate, "the four tigers" -- Taiwan, South Korea, Hongkong, Singapore -- and the emerging Chinese dragon. It is the capitalism that Eastern Europe and the splintering countries of the former Soviet Union want to emulate. All over the world, there has been a rush to dismantle ossified socialistic planning. As an empirical proposition, the gospel of Marx and Lenin and state socialism did not work.

I think it would be naive to believe that it means that capitalism has "won" over socialism. Americans like to see the course of events in terms of competing sports teams, "us" against "them," "the American way" against the "the evil empire." The capitalism that is now emerging as the prevailing economic order is not the capitalism of the Vanderbilts and J.P. Morgan. Precisely because capitalism generates change, the nature of capitalism itself changes. And the pace of change has been accelerating.

The emerging economic order is a new kind of capitalism that neither Adam Smith nor Karl Marx foresaw. There exists no accepted name of this new order. In this book I shall use the labels "creative capitalism" and "high tech capitalism." Those terms are tentative and may not be the terms that historians and economists will eventually settle for. But they do point at some key characteristics: the harnessing of man's creative urge to promote economic change, and the proliferating use of high technology (to be defined).

The economic history of the greater part of the twentieth century is the history of the competition between capitalism and socialism. It was a struggle that engaged social thinkers, politicians, soldiers dying in the battle fields, and common men and women. Two competing gospels were vying for the souls of the faithful, promising an earthly paradise. In the words of Arthur Koestler, socialism was the "God that failed."

Ideas are powerful. An idea can change the world. And yet, it was not the intellectual conception of capitalism -- however imaginative -- that eventually prevailed. Rather, it was the advent of a new kind of technology that turned the table.

Stumbling Giants and a New Breed of Startup Companies

The technological facts are not disputed. Technological progress before the 1880s was tied to steam power, railroads, steel. Another wave of innovations occurring around the turn of the century involved the chemical industry, electricity, and the automobile. Later there was aviation, petroleum, the movie industry, radio. Most of these "conventional" technologies (as we would say today) displayed marked economies of scale so that there were strong incentives for successful corporations to grow and to become very big.

Steel is a case in point. The Bessemer method for making steel revolutionized the iron and steel business in the early 1870s. Andrew Carnegie built the largest steel plant in the world in Pittsburgh, based on the new method. Carnegie kept the lead in the steel industry by regularly installing new and more efficient machinery and by taking advantage of new technological processes. In order to ensure a regular and plentiful supply of coke, he entered into a partnership with H.C. Frick. To ensure the supply of iron ore, he reached an agreement with Rockefeller, who owned large interests in central Minnesota and controlled the railroads and the Great Lakes steamships. As the Carnegie output of steel soared, the price of steel rails and other steel products fell.

Similar stories can be recounted for most major technological innovations that occurred during the nineteenth and the first half of the twentieth century. Technological progress went together with the accumulation of vast volumes of capital: automobile and aircraft plants, pipelines and refineries, electric utilities and high voltage transmission cables, and so on. The product was standardized and uniform (cf. Henry Ford's famous comment that his car could be obtained in any color as long as it was black) and it was possible to bring unit costs down by mass production. As prices were lowered, mass markets were created. Mass production and mass consumption.

In December 1900, Pierpont Morgan attended a famous dinner at the University Club in Manhattan. Morgan was seated next to the dinner speaker, Charles M. Schwab, the faithful lieutenant of Andrew Carnegie. Schwab invoked the vision of a steel trust that would control the world. A week later, Carnegie and Morgan shook hands: the U.S. Steel Corporation had been formed. It was the first billion dollar corporation in history. It held 75% of the American steel market. It was the apotheosis of monopoly capitalism.

Monopoly profit is a rent on the ability of management to eliminate its competitors. There existed a built-in incentive to destroy the competition and to wreck the rules of the market.

Then, in the 1970s -- or perhaps even earlier -- something happened, something that changed the rules of the game of capitalism. Technology itself changed. The large high-volume integrated plants of old with disjointed production lines are giving way to small plants with faster, cheaper production methods that can adjust quickly to demand variations.

The open-hearth furnace in the steel industry had long been obsolete. The atmospheric pollution was no longer deemed environmentally acceptable. The coke ovens and the blast furnaces will eventually disappear entirely. New technology permits the casting of thin slabs linked directly to a rolling mill in a continuous process. Today more than a quarter of the world's steel is made in electric furnaces. New "minimills" dominate the market for structural steel, and are challenging the market for sheet and plate steel. U.S. Steel (merged with Marathon Oil to form USX Corporation) and Bethlehem Steel are still the nation's two biggest steelmakers. But they are facing increasing pressure from new aggressive players in the market, like Nucor Corp. and Chaparral Steel running some of the world's lowest cost steel plants.

Under high tech capitalism, capital and profit is changing character. Capital is no longer just bulky integrated plants with big smokestacks. It is also, and more significantly, campus-like facilities manufacturing electronic chips, satellites placed in geo-synchronous orbit, and salmon farms. And it is not just hardware: it is blueprints for preprogrammed industrial robots, computer programs like CAD (computer aided design), and genetic codes in the pharmaceutical industry.

Monopoly profits and cartels are becoming rare as fierce international competition and the proliferating technical possibilities to develop alternative manufacturing processes make it difficult for any one producer to protect exclusive information and exclusive designs. In the high tech world there are always technological alternatives. The Swedish match king I. Kreuger built a world match monopoly in the early 1930s. It could not be done today, with cigarette lighters providing a cheap substitute. The Hunt brothers tried to corner the silver market. But their attempt was destined to fail as photographic film was rapidly losing ground to videorecording.

The only way to make a profit is to build a niche -- a product design niche or a technology niche or a marketing niche -- where the manufacturer can enjoy some limited brief exclusivity before the competitors catch up. Profits are doomed to dissipate if you do not launch immediate successor product designs on the market so that you can all the time stay one pace ahead of your competitors. Profits accumulate as the return on ongoing innovative product design and market development. Profit is a rent on corporate creativity.

As an example of what has now been said, consider the case of IBM. IBM never managed to create an enduring monopoly. At times it seemed to have the mainframe market sewn up, only to face increasing competition in peripherals and in the software market. IBM's position in the mainframe market was to be challenged as well, as other manufacturers pioneered the concept of the "supercomputer" (Cray and Amdahl). The market for mainframes has been eroding for some time anyhow, as new generations of computer "workstations" (pioneered by Apollo and Sun) and new, incredibly powerful PCs have come on line. To survive as a computer manufacturing giant, IBM must go after an entire array of market niches.

In 1992 IBM suffered a record loss of almost 5 billion dollars, on total sales of $ 64 billion. The price of IBM stock collapsed. IBM laid off employees for the first time in half a century. The departing chief John Akers speeded up plans to dissolve the monolithic structure of the corporation, vowing to break up the company into smaller autonomous units. But the new boss, Louis Gerstner (hired from RJR Nabisco) decided to go slow, embarking upon a program of cost-cutting and streamlining the international organization of the company.

Is Microsoft Corporation a monopoly? The rapid ascendancy of Microsoft and the plight of IBM illustrates the dramatic turbulence of the high technology world. The stock-market value of Microsoft now tops that of IBM. Microsoft's MS-DOS operating system dominates the software market. But after a year-long investigation by the Justice Department, Microsoft agreed to make some minor adjustments in its marketing practices and was let off the hook. Software is not a static product. It evolves, in new applications and in response to the development of new hardware. The competition includes both established competitors like Lotus and Borland and hundreds of small software companies, all nibbling at the market, searching for new approaches and new openings.

The new catchword in the management of large U.S. corporations is "outsourcing." It is the very opposite of vertical integration. General Motors is outsourcing the manufacture of thousands of parts and components of its automobiles, buying them in the marketplace rather than manufacturing them. Why? Because small firms, specializing in niches of technology, are often able to develop and manufacture more advanced products than a mega-company.

The large companies are certainly not going to disappear. Many of them will survive, getting "leaner" and "meaner," trimming in-house operations to their core competencies and farming out other work to rings of outside suppliers.

Economic Diversity

By an ironic twist of fate, it took a Marxist to discover that the prevalent market form under capitalism was no longer monopoly but product development and competition between name brands. To describe that market system, Joan Robinson in 1933 proposed the term "imperfect competition." In economic textbooks it is known as monopolistic competition.

Under monopolistic competition, producers need to introduce a continuous stream of new product designs. The result is that there exist few uniform products any longer. A product is designed and marketed to meet the approval of a particular target customer group. It has a brand name and distinctive technology that is recognizable to the customer. It is also individualized in the sense that there is no other product that is quite the same.

Modern products have many attributes and many uses. They are multi-dimensional. Each product delivers a bundle of services rather than covering a distinct single need. A breakfast food will provide you with calories, taste, fiber, vitamins. Perhaps it even reduces your cholesterol level. A personal computer can be described by its computing power, its disk drives, the resolution of its graphics, the software it supports, and its networking capabilities. A fiber optic cable delivers speed, clarity of messages, and amplification. A CATSCAN machine in a hospital has attributes such as resolution, penetration and versatility. There is a sort of double multiplicity in the marketplace in that each product delivers a bundle of individual services, and that each desired service can be obtained by an array of alternative competing name brands.

Competition in the marketplace is foremost in terms of quality, i.e. in terms of the bundle of attributes that each brand name delivers. Each brand has its own market niche. There is no single mass market, but a great number of market segments. It is not obvious that there are any pronounced economies of scale. Small firms and large firms exist side by side in the market. Product development leads to further fragmentation of markets. It leads to a wider choice for the consumer.

It is not just consumer goods that have become multidimensional. The same goes for capital goods. A welding robot, a stacking crane, a computerized inventory system, an operating room in a hospital all are able to deliver an entire menu of production services. There is all the time a quest for delivering more quality, not just quantity.

The explosion in economic diversity that occurred during the latter part of this century actually arrived quite unexpectedly -- something that economists in general did not foresee, and something that economic theory even today is poorly equipped to handle. Alvin Toffler, in his bestselling book *Future Shock* (1970), saw these things clearer than many others. As technology becomes more sophisticated, the cost of introducing variations declines. Markets become segmented or even fragmented into a host of mini-markets, catering to the increasing diversity of the consuming public. In the fifties and sixties, youngsters and teenagers would go to see a drive-in movie on a balmy summer night. Where are all the drive-in movie theaters gone? The old-style mass cinema audience has disappeared. Instead, multiple small audiences turn out for particular kinds of films. Most movie theaters have been converted to clusters of four or six showrooms. Toffler found similar examples in the U.S. automobile industry (Ford customers designing and ordering their own personalized Mustangs), and in publication (magazines aimed at surfers, scuba-divers, hot-rodders, skiers and jet passengers). Toffler discussed the drive toward ethnic diversity and "multi-culturalism" in the U.S. 25 years before it became a hot social agenda. His theme: mass markets are being "de-massified," moving from homogeneity to heterogeneity.

The U.S. economy is rapidly becoming transformed into a "quality economy" rather than a "quantity economy." To be able to sell a product, it is more important to deliver quality than quantity. Eventually -- presumably some time next century -- all U.S. families will have raised their standard of living to the point where they already own the cars, TV sets, refrigerators and jaccuzzis that they want. You can only have so many cars parked in your driveway. At that point most sales in the country will be replacement sales. The national markets will stagnate in terms of quantities. But there will still be scope for technological progress, of course -- progress in terms of better qualities rather than greater quantities.

In theory, improved qualities should be reflected by an increase in the gross national product, or GNP. But it is notoriously difficult to measure statistically the value of improved quality. It is generally agreed that the official statistics seriously underestimate the value of improved goods and services. That may be one reason why the GNP figures lately have had a rather poor showing in this country. Frozen food is so much better now than twenty years ago. A dollar's worth of transportation services will bring you to your destination faster and more comfortably. The sound from a CD player is better than the sound from an old-fashioned grammophone turntable. The advice and the procedures of your doctor are more helpful. The stagnating output figures during the 1970s, and the feeble growth in the 1980s, no doubt reflect the fact that the U.S. economy is spending more effort on delivering diversity and quality rather than just more quantity.

In the face of the new diversity of the U.S. economy and the disappearance of the mass markets, much of the accepted economic science needs to be rewritten. One economist who has grasped the dimensions of this task is Robert Reich, the Harvard economist whom President Clinton appointed as his Secretary of Labor. In his book *The Work of Nations,* Reich explains the transition from "high volume" to "high value." The manufacture of uniform and standardized products is a thing of the past. Instead, successful businesses are moving to specially tailored products and services, finding the right fit between niche technologies and niche markets.

What is High Technology?

During the twentieth century, capitalism has evolved into high technology capitalism. I shall define high tech this way: it is the technology of long and hierarchical production chains delivering diversified products or services with many attributes.

A motor vehicle is assembled from thousands of individual parts. Each component has its own manufacturing and assembly history. The gear box was built separately. The fuel injection system uses electronic chips to monitor the performance of the engine. The chips were manufactured by a semiconductor company. The brakes, the airbag, the audio system all have their own separate technologies and sub-technologies. The manufacturing architecture of the final product is hierarchical, a network of individual production links. To use a term that I shall expand on later in this book, the entire manufacturing process can be viewed as a "logistic system."

As capitalism has progressed, this hierarchy has become more complex. The chains of individual links have become longer. Also each link, be it a manufacturing link or a distribution or a marketing link, has become more diversified. Product development upgrades or adds dimensions to an already multidimensional product. Product development in the home security industry is no longer concerned with just monitors and alarm sirens and patrol cars. New features that are being added these days are hookups to emergency medical services, electronic identification of visitors, and a variety of services for the elderly. The "smart" credit card (which contains information about the account holder and current account balances in a microchip on the card) adds new financial services to the user of the card. Product development by genetic engineering in agriculture aims at developing new strains of wheat and other cash crops that will tolerate greater variation in temperature and moisture, that are resistant to pests, and that grow faster. Product development in medical technology such as magnetic resonance imaging has opened up new possibilities for doctors to get a look inside the human body.

Furthermore, high technology is all the time becoming "higher," i.e. the logistic system grows hierarchically with new and even more diversified technology layers added to the earlier ones. In the vivid account by J. Schumpeter (1883 - 1950), innovating firms searching to acquire a competitive edge are seen as the catalysts of the capitalistic process, pushing the technology frontier ahead of them. The kind of innovations that Schumpeter had in mind were discrete jumps in technology, like the automobile, the airplane, the telephone, synthetic rubber. Such bursts of evolutionary creativity still occur (telefax, cellular telephone, communications satellites, open heart surgery). Often they are accompanied by massive extinction events, as the earlier technology becomes obsolete (nobody sends telegrams anymore, and Western Union is fighting oblivion).

But the majority of changes of technology and of products occur as the cumulative sum of a large number of small changes of product design rather than big discrete jumps. Most innovations are no longer the result of one inspired genius (an Edison or a Nobel) or a farsighted entrepreneur (Ford) but are planned and subject to management and control. In the intensely competitive climate of the 1990s, a firm needs to bring to the marketplace a continuous stream of new product designs, each generation being superior to the preceding one. Consider the cases of the modern kitchen machine, or the garment industry, or even the automobile. New designs hit the market every month. But the change from one "vintage" to the next is minute.

One characteristic of high technology is its "complexity." This is the watchword of modern nonlinear dynamics. The Santa Fe Institute, an offshoot of the Los Alamos National Laboratory which I shall have more to say about later, publishes a journal titled *Complexity*. All living systems are complex. The economy, too, is a living system, with all the spontaneity and complexity of molecular biology. DNA or the genetic blueprint is a kind of molecular-scale computer that directs how the cell is to build itself and repair itself and interact with the outside world. In a similar manner, technologies are blueprints for the production of economic goods and services. A technology can be thought of as a kind of computer that directs the organization of a particular industrial or marketing activity. It includes instructions on how the technology can interact with other technologies, how it can be managed, and how it can be developed under the aegis of various corporate entities: how it can evolve.

Technological Evolution at the Edge of Chaos

Manufacturing technology is not alone in growing in a hierarchical fashion, with new stages added to the former ones both vertically and horizontally. Nature grows in the same way. The evolution of plants and animals is also hierarchical. The step from ape to man was a case of "high" genetic technology becoming even higher. The end product -- man -- is the result of a longer evolutionary chain. He is also more versatile and adaptable than the ape.

In a way, the modern conception of evolution is still Darwinian. The survival of the fittest is still seen as a driving mechanism. But there is more. During the last thirty years, a large body of knowledge has been brought together concerning the evolution of dynamic systems in general. The breakthrough occurred in thermodynamics, in physics, and in meteorology. At its core, this theory is mathematical, dealing with the behavior of "nonlinear" systems, like water dripping from a faucet or the rising morning mist. This is the modern theory of chaos.

The basic premise of chaos theory is the universal tendency toward disorder, dissolution, and decay. The rise and the fall, the bankruptcy, and the death of corporations. Mergers and acquisitions, corporate raiders and corporate vultures. Be it a piece of machinery, a living organism, or a corporation : for a brief moment, life is a triumph over death. But the triumph is forever evanescent. Life has to be reestablished every second.

Complex systems have at their heart a great number of "agents" such as molecules or neurons or species. In economic systems the agents are managers and corporations. To fight the cosmic compulsion for disorder, agents are constantly organizing and reorganizing themselves into larger structures through the clash of mutual accommodation and mutual rivalry. Molecules form cells, neurons form brains, species form ecosystems, corporations form industries. At each level, entirely new properties appear. And at each stage, entirely new laws, concepts, and generalizations are necessary. Each system has many niches. The very act of filling one niche opens up more niches. The system is always unfolding, always in transition. New technologies and startup companies representing new ways of doing things are forever nibbling away at the edges of status quo, and even the most entrenched old production methods eventually have to give way.

Emergent technologies are those that are rising above the horizon of technological and commercial feasibility. Often there is a competitive race among many companies seeking to exploit a new technology, such as the race among computer companies a couple of years ago to develop the laptop computer technology, or the current race to develop high-definition television. Emergent industrial structures are new ways of organizing R&D, production, distribution and/or marketing. Often new technologies prompt dramatic changes of industrial organization (examples: advances in tele-communication make it possible to have employees telecommuting, genetic improvements of fruit permits longer storage and distribution over larger geographical areas, new pipeline technology makes it possible to drill oil on Alaska's North Slope and to distribute it world-wide).

Perhaps the most important characteristic of emergent technologies is the inherent risk and uncertainty -- the risk that the technology won't work, and the uncertainty how the components of the new technology are going to fit together. Consider "multimedia" -- the marriage of computers, video, TV, and audio. Obviously it is going to come. Obviously it will entail an electronic revolution. Billions of dollars will be made. But what is it going to look like precisely? Is it going to be transmitted via optic fibers of the cable networks, or is it going to be broadcast over the airwaves? And which are the corporations that stand to benefit? Cable companies or Hollywood? Software companies like Microsoft or the manufacturers of "smart" boxes to be put on top of all the TV sets in the land? A new industrial structure will emerge. In retrospect, it will no doubt look perfectly compelling, orderly, and logical.

To understand the nature of the emerging industrial state, consider the computer superstars of the 1980s that humbled the once-mighty IBM. They were all startups like Apple, Sun Microsystems, Compaq, Silicon Graphics, and Dell. The workstations and personal computers that they pioneered were more than just new computer products that reordered the product mix of the computer industry -- these new products wrought fundamental changes in the economics and the structure of the computer industry. These companies maintained low overheads, and they outsourced many components to save R&D expenses. Smaller companies like Cyrix, MIPS and Tseng Labs could be viewed as design houses or design boutiques farming out the manufacturing to outside companies.

The high technology race requires a new organizational mindset and a new corporate culture. It includes respect for individuals and their ideas, and an open and pervasive communication system throughout the corporation. The 3M Corporation challenges its various business units to have 25 percent of their sales in any given year from products that are less than five years old. Any employee who has a product idea is allowed to spend 15 percent of his or her time pursuing it. The company has a dual technical/management ladder system that recognizes and compensates research excellence and contributions on an equal basis with corresponding management positions.

The winds of change are sweeping new industries and old industries alike. To prosper, a company needs to bring a continuous flow of new product designs successfully to the market ("commercialization"). The most significant competitive weapon is time to market. The management of product development is an instance of planning under uncertainty. The managers must develop plans for handling various contingencies; they cannot place all eggs in one basket. The new industrial climate requires rapid decision making, a lean and horizontal organizational structure, and a willingness to take risks. The optimal innovation policy is to pursue an entire portfolio of technological development projects. Many of these will fail the test of commercialization, but a few will survive. Those that do survive will have only a limited life span, and will soon need to be replaced by superior products.

What will this never-ending quest for technological innovation bring us? The surprising answer of modern nonlinear mathematics is that the tendencies to order and chaos -- to build and to destroy, to invent the new and to discard the old -- somehow can keep each other in balance, never locking into a fixed pattern nor ever dissolving into uncontrolled turbulence. That balance is called "the edge of chaos." At the edge, the system has enough stability to sustain itself and enough creativity to evolve.

Furthermore, at that edge there is "self-organization." It is an aspect of evolution that not even Darwin understood. It goes beyond the principle of survival of the fittest. As a storm picks up speed it may evolve into a hurricane. The eye of the hurricane travels along some path, from the Bahamas toward Florida. Eventually the hurricane dissolves. Migrating birds develop the ability to find their way over half the globe. Land-based animals develop organs like the lungs. General Dynamics develops into one of the major defense contractors in the U.S., and eventually, in 1992, a decision is made to sell out most of its divisions and to redistribute the cash proceeds to the stock holders. Structures form spontaneously as order battles chaos.

Evolution never proceeds at an even and predictable pace. There are bursts of creativity interspersed with long periods of inactivity, or even collapse. One such burst of technological creativity occurred in the early 19th century -- the industrial revolution. To use the terms of dynamic theory, it represented a "phase transition." Are we now, in the early 1990s, heading toward yet another such economic metamorphosis, a phase transition to an age of instant communications, robotics and biotech? The transient state would be characterized by order and chaos intertwined in an everchanging dance of evanescent corporate structures. Eventually, from the shake-out, a new state of capitalism would arise.

Modern capitalism is in a perennial state of disequilibrium: a gap between on the one hand the perceived technological potential of the future and on the other existing practices that are fast becoming obsolete. It is a neverending quest toward technological targets and market potentials that are forever evolving. Capitalism is the management of the future itself.

The Collision with the Environment

The change-over from an earlier economy based on mass production of standardized and uniform products to the cornucopia of high technology has proceeded unevenly, and there are still large sectors of the U.S. economy based on the technology of an earlier age. Foremost among them are agriculture, forestry, and the fisheries -- all industries of resource extraction. Seen from the point of view of conventional technology, these industries still seem to offer strong economies of scale. The result: the disappearance of the family farm, big-corporation farming, clearcutting in the nation's forests, overfishing of the fishing grounds. Technology and the environment are both living and evolving systems; if one grows topsy-turvey, it will damage the living space of the other. And so, in the latter half of this century, the two systems collided head-on .

Classical economists used the term "land" as a shorthand for all natural resources, like farmland, forests, mines, waterfalls, rivers, and lakes. They viewed the sum total of these natural resources as an "endowment," a kind of original capital in kind that the good Lord had created for man. Economics used to be defined as the science of how it is possible to satisfy a given set of consumer wants with the products that can be obtained from a given endowment of resources. That has all changed. Man can produce his own "land" and his own natural resources. We can claim land from the sea (as in the Netherlands) or from the desert (Israel). Through the use of fertilizer we can convert low-yielding soils into high-yielding soils. We can manufacture drinking water (the desalination plant located at Ash Shuwaykh in Kuwait produced 75 million gallons a day before the Gulf war). In principle, even the supplies of fossil fuels can be replenished, e.g. by manufacturing synthetic gasoline or other synthetic fuels (alcohol, for instance). And the technology of locating and tapping existing deposits of fossil fuels has made quantum leaps (seismic exploration, offshore drilling, and drilling in the Arctic.)

The most dramatic advances in creating new natural resources are occurring in agriculture and in biotechnology. The so-called green revolution has produced new strains of plants, like hybrid rice. Modern biotechnology is creating new strains of grains, vegetables and fruit. Through marine biotechnology we can grow salmon, crab, crayfish, and even cod in large nurseries. Scientists these days can create new cattle and new poultry that are disease resistant and deliver meat of desired quality. The total supply of resources is no longer a given datum. Instead, the supply of resources itself can be managed and be made to grow and develop.

The results have certainly been impressive. Global food production is expanding. World grain output rose about 3 per cent per year in the time period 1950-84 (but less rapidly more recently). Nevertheless, the global population figure is increasing even faster, so that per capita food production has been and is falling in many parts of the world. The outlook for the immediate future is uncertain. The supply of available cropland is rapidly diminishing due to soil erosion, degradation, and to make way for roads and buildings. The rainforests deserve to be protected, not denuded.

Even advanced agrotechnology often relies on a plentyful supply of some natural resource. The Salinas Valley south of San Francisco is home to some of the most advanced agriculture in the world; yet, its future is clouded as the groundwater recedes (due to over-irrigation) and saltwater from the Pacific seeps into the aquifer.

The most striking example of a commodity that is still produced under conditions of vast economies of scale and which is perfectly standardized and uniform is energy. One kilowatthour looks exactly like another. With an insatiable appetite for more energy, the modern economy has rushed into a head-on collision with the environment, damaging the air, the lakes (acid rain), and even the climate (the greenhouse effect).

Considering the vast destruction of the environment, the growth in GNP that has been experienced by many developed and developing nations during the second half of the twentieth century may have been largely a sham. We do not yet know the full extent of the damage. But that is precisely the point. These damages have occurred because we were taken unawares and because there was lacking scientific knowledge about the effects of pesticides and herbicides on the ground water, the effects of chlorofluorcarbons on the ozone layer surrounding the globe, and the long-run effects of the burning of tropical rain forest. As these matters are being subjected to scientific and public scrutiny and as our understanding of them improves, options of alternative technologies present themselves.

Restoring the environment and safeguarding it offers tremendous technological challenge. The challenge itself is a powerful stimulus. Man built vessels and boats in order to traverse the sea. He built the Apollo to go to the moon. In a society that is willing to accord environmental repair work a high priority, there will also emerge incentives to develop new environmental technology. There is money to be made from converting waste to energy, from building refrigerators using helium or natural gas as coolants, from solar energy and wind energy and wave energy, from cleaning up the beaches and designing and building attractive resorts, and from producing and selling organic food. The capitalistic system is not anti-environment. It is pro-change. Capitalism is now being harnessed to correct the damage that has been done.

Toward the middle of the next century, small may become beautiful also in agriculture and energy extraction. One can see the beginnings of this development already today in organic farming. And it will happen on a large scale when grain and fruits can be grown *in vitro*, in the laboratory, rather than on huge farms. Researchers already grow cellulose fibrils and paper in the lab (see the box "A Visit to Malcolm Brown's Laboratory" later in this book). When such technologies hit the marketplace there will be an explosion of diversity, with consumers being offered synthetic grapefruit pulp of any imaginable color and taste, or synthetic cotton textiles of any imaginable texture or strength (all controlled at the molecular level by appropriate biotechnology). Present farmland and logging land will be returned to its pristine status, set aside for recreation, wildlife, or just for its natural beauty.

Where Did Marx Go Wrong?

Let me first point out where Marx was right. Marx argued that social structures were transient historic forms determined by the productive forces. In *Das Kapital* (published in 1867), he argued that the development of capitalism is accompanied by increasing contradictions. It is subject to internal pressures resulting from its own development. He predicted that capitalism would be shaken by ever more severe crises. Marx was the first "chaos economist" -- one hundred years before mathematicians had coined the concept of chaos. His big mistake was to believe that the underlying forces of technological change could be tamed by the overthrow of the existing social and political superstructure, replacing it by the dictatorship of the proletariat.

Stalin instituted rigid planning directed from a national economic planning center (the "Gosplan"), nationalization of all farms and all industry, and total control of both thoughts and actions of all Soviet citizens. Such centralized control had become possible in the 1920s because the information technology of the day permitted it: the control of the press and radio, a secret police with vast resources, and concentration camps (the Gulag). A hierarchical centralized information and command system was set up, channelling all information to the top, with the directives flowing in the opposite direction.

With sufficient exertion, the totalitarian state can crank up the economy to manufacture increased volumes of everything that is uniform, standard and routine. It can build big steel mills (Magnotogorsk) and big hydroelectric power stations (Volgograd). These were the dinosaurs of an earlier technological age.

The fascist states of Mussolini and Hitler were based on the same principles. The economic reality was the same: rigid centrally directed production hierarchies, and strong economies of scale manufacturing standardized goods. It should be noted that this kind of economy was not without success. A nation that was willing to pay the political and cultural and moral price could in this manner accelerate its economy and become fully employed, with a rapidly rising GNP. The problem with state socialism is that it cannot generate internally the technological change that satisfies the needs and the demands of the consumers. It does not provide for the diversity in the marketplace that alone can trigger the evolution of high-value-added and knowledge-intensive goods like computers, communications, and pharmaceuticals. The Soviet Union had fundamental problems already at the point of implementing conventional and known technologies. It was not able to provide its people with clothes and shoes and automobiles. It was entirely unable to develop new technology that met with the approval of its consumers.

There are two aspects to this question. First, in order to push innovations that benefit the consumer, you have to perceive what the consumer needs. In brief: marketing. There was no marketing in the Soviet Union. The collective society promotes conformity. Conformity is the very antithesis of creativity. Marketing means to search out the preferences and tastes of the individual consumer. If the market is uniform, there is no need and no scope for marketing. The economic planning directorate determined what kind of shirts and shoes the Soviet consumer should be able to buy in the store. The producer determined for the consumer.

Second, you need to develop feasible technology that can satisfy the perceived needs. This second part is engineering. In *The First Circle*, Solzhenitsyn tells the story of a team of physicists, radio engineers, and mathematicians, all recruited from Stalin's concentration camps, commandeered to develop new electronic eavesdropping equipment. (The "first circle" was the privileged place in Dante's Hell reserved for wise pagans.) The researchers put in grueling work weeks, exerting themselves to the limit of their abilities, in constant fear. But in the end, they failed to meet Stalin's demands, some were executed and others were returned to the labor camps. Hitler's rocket scientists at Peenemunde were decades ahead of their time, working in teams with military discipline, under fear of repression. Hitler's aviation experts even built a jet engine fighter airplane. But the Fuehrer did not like it and the project was shelved. Technological progress requires a plurality of decentralized decisions. It requires competition.

State socialism reached its apex in the 1950s and 1960s. Economic growth in the Soviet Union was still respectable, maybe around 5 % per year. The new nation states in Africa and Asia, liberated from colonialism, looked to the Soviet Union for economic guidance. But from there on it was all downhill. As the new decentralized knowledge-based technology entered the industrial scene in the West, the Soviet economy was frozen in a straightjacket of planning and regimentation.

The dreamers of socialism never understood the capitalist process of material creation. The government cannot create; it can administer. Creativity requires freedom of the human mind, the soaring of the human spirit. Creativity is a many-splendored thing. No two individuals dream the same dreams, see the same visions. Artistic advance and material advance arise as man tests and probes the quality of the new and discards the old. It cannot be accomplished by groupthink. It cannot be accomplished by people in bondage. It requires competition of dreams and competition of ideas.

The collapse of state socialism was swift. A free fall of the economy, a self-reinforcing downward spiral. The rigid hierarchies that could not accommodate change had to give way. The utter irony of it all was that the cataclysm that Marx once had predicted to befall capitalism instead engulfed the dictatorship of the proletariat.

The Freedom to Create

The relationship between capitalism and freedom is of course a favorite theme among political writers. To many people, capitalism and liberalism are synonymous. The issue needs to be reexamined in the light of the emerging new order that I propose to call high tech capitalism. There are two kinds of economic freedom that must be present for an economy to be able to pursue a path of creativity: freedom of market formation, and freedom of entrepreneurs to be creative. Free markets function like a computer or a "magic hand" that solves the basic problem of allocation and production and distribution in an economy. I do not think that I need to pursue these points. But the second kind of freedom has attracted less attention by professional economists and is not commonly well understood. It is the freedom to "tinker," the freedom to experiment, develop, and install new technology. It is the freedom to change things. It is the freedom to consider alternatives.

Creative freedom does not mean absence of rules. To listen to the Brandenburger concertos by J.S. Bach is to be transported to a world of true human inventiveness, where musical caprices adorn and complement the magistral progress of the thematical development. And yet, this music is nothing but pure mathematics, constructed according to rigid rules of harmony and contrapunct. The true genius reflects itself in its ability to fill rules by spirit. In a very similar manner, an inventor like Thomas Edison was able to pour his genius into the rigid laws of electricity and mechanics and to unlock the soaring possibilities -- and the caprices -- of the new communication age.

Creative freedom does not require *laissez faire*. It can be managed and it should be managed. The perspective that mankind would not and should not be able to influence the path of technological progress would be a horrifying one. That would leave us with no way of influencing our own future. It would put us in the roles of actors in a Greek drama, vainly trying to stake our own course but unwittingly complying with the predetermined design of Olympian gods. There are technological alternatives, for single corporations and for the entire nation. There could have been alternative pasts. There exist alternative futures. We have a choice and we need to exercise it.

This book traces the technological development and potential in communication, electronics, robotics, biotechnology, and energy generation. It surveys recent innovations in equity markets and in financial markets. It explores the creative power that propels such advance. It also probes the directions where we are headed and the responsibility of enlightened government to set sign posts for the road ahead.

Bibliographic Notes

One of the leading ideas of this book is that technology during the 20th century changed from the mass production of uniform products to the development and commercialization of continually upgraded technology "generations" delivering diversified products, and that this change has fundamentally rewritten the laws of capitalism.

The analysis of consumer goods in terms of attributes builds on K. Lancaster, *Consumer Demand, A New Approach*, Columbia University Press, New York 1971. But whereas Lancaster described a consumer good as a vector of attributes, I prefer to see it as a "tree" of attributes. Over time, the attribute tree can grow in two ways: existing attributes can be upgraded, and entirely new attributes can be added into the preference structure of the consumer. (Cf. also the theoretical developments of a "utility tree" in W.M. Gorman, "Separable Utility and Aggregation," *Econometrica*, Vol. 27, July 1959, pp. 469-481.)

Drawing on Lancaster, M. Trajtenberg has developed an approach to product innovation based on the idea of "product classes" evolving over time. The different brands of each class can be described in terms of a small number of attributes and prices. Product innovation constitutes changes over time in the set of available products. See M. Trajtenberg, *Economic Analysis of Product Innovation -- the Case of CT Scanners*, Harvard Univ. Press, Cambridge, Mass. 1990.

Management textbooks have of course long recognized the differentiation of products in markets; modern texts also give advice on developing suitable corporate differentiation strategy. In his two bestselling books *Competitive Strategy: Techniques for Analyzing Industries and Competitors*, The Free Press, New York 1980 and *Competitive Advantage: Creating and Sustaining Superior Performance*, 1985, Michael E. Porter discusses various "uniqueness drivers" open to the firm; they include not only distinctive engineering and marketing of a product but also individualized attention to personnel, materials handling, transportation, manufacturing, sales policies, and servicing -- that is, to the entire "value chain" of a product.

A good account of current changes on the American industrial scene is given in R.B. Reich, *The Work of Nations: Preparing Ourselves for 21st Century Capitalism*, Knopf, New York 1991. Reich gives many examples of the changes that are occurring in the manufacturing industries, all involving shifts from the uniform, the routine, and the standard, to customized products. But Reich goes barely beyond the recording of such observations. The theory of proliferation of attributes digs deeper, providing an explanation for what is happening.

Economists agree that the improvement of quality (of consumer goods and of capital goods) poses formidable problems both in theory and in statistical estimation. For a discussion of the problem as it relates the construction of so-called hedonic price indices for computers and computer services, see Cole, R., Y.C. Chen, J.A. Barguin-Stolleman, E. Dulberger, N. Helvacian and J.H. Hodge, "Quality-adjusted price indexes for computer processors and selected peripheral equipment," *Survey of Current Business* 66, 41-50, 1986. In the text, I am conjuring up a future economy where all growth would take the form of improved quality. To my knowledge, no formal analysis of such an economy has ever been attempted.

References

"In the words of Arthur Koestler..." Arthur Koestler and others, *The God That Failed*, Harper, New York, 1st ed. 1950.

"A. Carnegie built the largest steel plant in the world..." M. Josephson, *The Robber Barons: The Great American Capitalists, 1861-1901*, Harcourt, Brace& World, Inc., New York 1934. (Paperback edition in 1962.) Also F. W. Boardman, Jr., America and the Robber Barons, 1865 to 1913, Henry Z. Walck, Inc., New York 1979. One would think that scholars by now should have arrived at some kind of consensus regarding the unprecendented economic boom of late 19th century America. But they have not. Josephson's book with its colorful gallery of villains became an instant classic when it was published. It still enjoys a popularity out of all proportion of the book's merit, and in spite of much later scholarship discrediting many of Josephson's sweeping assertions.

"In December 19901, Pierpont Morgan..." R. Chernow, *The House of Morgan*, Atlantic Monthly Press, New York 1990, p. 85.

"New 'minimills' dominate the market..." D. Milbank, "Big Steel is Threatened by Low-cost Rivals, even in Japan, Korea," *Wall Street Journal*, Feb.2, 1993.

"Monopoly profits and cartels are becoming rare..." Antitrust legislation served a useful purpose in an earlier era, but it is rapidly becoming redundant in the high tech world. The last major piece of antitrust legislation in the U.S. was the breakup of AT&T in 1984.

"...consider the case of IBM." R. Sobel, *IBM: Colossus in Transition*, Truman Talley Books, New York 1981.

"The large companies are certainly not going to disappear. Many of them will survive ..." B.Harrison, *Lean and Mean: The Changing Landscape of Corporate Power in the Age of Flexibility*, Basic Books, New York 1994.

"By an ironic twist of fate, it took a Marxist..." J. Robinson, *Economics of Imperfect Competition*, McMillan, London 1933. See also E.H. Chamberlin, *Theory of Monopolistic Competition*, Harvard University Press, Harvard, 1933, 2nd ed. 1936.

"Alvin Toffler, in his bestselling book..." A.Toffler, *Future Shock*, Random House ,New York 1970.

"In the vivid account by J. Schumpeter..." J.A. Schumpeter, *Business Cycles*, Vols. 1-2, McGraw-Hill Company, New York 1939 and J.A. Schumpeter, *Capitalism, Socialism and Democracy*, Harper and Brothers, New York, first edition 1942, subsequent editions 1947 and 1950.

"This is the modern theory of chaos..." M.M. Waldrop, *Complexity: The Emerging Science at the Edge of Order and Chaos*, Simon & Schuster, New York 1993, and R. Lewin, *Complexity: Life at the Edge of Chaos*, Macmillan, New York 1993. See also the main text, "The Long Waves."

"The 3M Corporation challenges its various business units..." J.A. Woolley, "Commercializing New Materials: A 3M Perspective," in ref. [12]. (For numbered references, see the list at the end of the book.)

"World grain output rose about 3 per cent per year in the time period 1950-84..." L.R. Brown, State of the World 1990, New York. Cited after P. Kennedy, *Preparing for The Twenty-First Century*, Random House, New York 1993.

"In The First Circle, A. Solzhenitsyn..." A.L. Solzhenitsyn, *The First Circle*, Harper and Row, New York 1968.

"Economic growth in the Soviet Union was still respectable..." J. Cracraft, *The Soviet Union Today: An Interpretive Guide*, Chicago, 1988.

CHAPTER 2

New Webs and New Constituencies

As the diversity in the marketplace is exploding, many managers ask themselves how they shall be able to maintain the present hectic pace, churning out a never-ending series of new technology generations. The product cycles are getting shorter and shorter. Is there no end?

The answer is that new hierarchies are being established, taking the place of the discarded vertical production alignments of an earlier industrial era. New webs of capitalism are being spun. New constituencies arise.

These hierarchies are the children of the electronic age: networks between corporations and corporate officials that span the nation and the globe. Loosely organized networks of pooled information, shared research and development efforts, strategic alliances, management coalitions that are quickly set up and quickly fade again as they are replaced by new and superimposed patterns .

The aim of such collaborative effort is synergy -- or, to use a simpler word, piggy-backing. Each new technology piggy-backs on its predecessor. Working together, corporations are searching to find synergy from each other, piggy-backing on their common efforts to develop the next generation of technology. And since all future technology is uncertain, a corporation has a strong incentive to reach out for that technology in many directions, to many constituencies. There are economies of scale in research and development that benefit the entire technological community. To tap into this synergy, a corporation has to pool its research efforts with others.

Peter Drucker calls the emerging form of business society the "network society." Economists working at the Santa Fe Institute call it a "web" (as in the World Wide Web of the Internet). Whatever term we choose to use, the industrial organization of the future will clearly be much more loosely and more temporarily laid out than the rigid hierarchies of the past. There are strong centrifugal forces present in these temporary organizational forms, where a company, hospital , or government agency turns over an entire activity to an independent firm specializing in that kind of work. Outsourcing information systems, purchasing and maintenance functions are becoming common already today. Perhaps the corporation of the future will become just a shell ("the virtual corporation") offering career opportunities for its senior management but delegating all its support functions to the outside. These virtual corporations will be amorphous, reaching out to other corporations in an ever-changing dance of cooperative arrangements, strategic alliances, and partnerships.

What is a Strategic Alliance?

In 1991, Apple struck a watershed agreement with rival IBM. The world's two biggest makers of personal computers entered a joint venture to develop future system software and the next generation of chips. The pact was a pre-competitive alliance for research and development. Apple later this decade will need a successor to the company's Macintosh computer line. The Macintosh is not compatible with MS-DOS, the PC standard supplied by Microsoft. For some time, Apple has found it difficult to sell its incompatible PCs to large corporate customers. Together with Big Blue, Apple will now have a chance to set the software standards of the future. The new "Taligent" operating system is scheduled for delivery in 1995.

In another collaborative initiative, Motorola in partnership with Apple and IBM developed a new powerful computer chip, the PowerPC microprocessor. It competes head-on with Intel's Pentium chip, and the developers hope that it will eventually break Intel's lock on the market (currently, the Intel family of chips hold roughly 85% of the PC market).

U.S. corporations have entered into thousands of research coalitions with other partners, both domestic and foreign. The coalition will stick as long as it is useful to all participating partners. But there are no allegiances beyond a company's own self-interest. For a while, Microsoft and IBM had entered into a strategic alliance with Microsoft providing IBM (and all IBM clones) with its MS-DOS operating system. But eventually the partnership fell asunder. Microsoft threw all its resources behind its immensely popular Windows software system. It withdrew its support from OS/2, a rival approach to PC software that Microsoft had created with IBM's help beginning in 1986.

Many U.S. high tech companies are striking up strategic alliances with Japanese partners. Apple computer is working with Sharp Corp. to make portable penbased computers. Aircraft manufacturer McDonnell Douglas and Fujitsu Ltd. have announced a wide-ranging factory-automation alliance, pooling U.S. software and Japanese manufacturing expertise.

The culture of many U.S. companies reflects a mentality of macho "go-it-alone." An American culture of broad corporate collaboration would be something new. Yet, the transition has to be made if the U.S. is to remain in the vanguard of high tech. This new corporate culture is forced upon us by the changing nature of technology. The technological uncertainty and the market uncertainty is breathtaking. Each corporation has its relative strengths and its relative weaknesses. By pooling their resources of research and product development, two corporations can create a temporary match powerful enough to unlock the next step in the accelerating technological race.

How is it possible for two companies to compete and collaborate at the same time? Once the technical problems at hand have been solved, each company will market its own product and aim for its own maximal market share. That is, as each new technology evolves along its life cycle, it will typically also move from collaboration to competition. Technologies and products that are still in the R&D pipeline, awaiting commercialization, are often the subjects of collaborative efforts. Once they are commercially viable, they have to make it on their own. To manage these transitions, a company has to conduct a flexible policy of entering into coalitions of cooperation with some partners, while perhaps letting other coalitions break up. The management of this "foreign policy" of a high tech corporation requires consummate skill and diplomatic tact.

The Japanese have developed institutional formats of collaboration that seem to point the way to the future. *Keiretsu* is a peculiar form for company coalitions, families of subcontractors who supply assembly plants with parts. The result is a production system that is distributed among many firms but yet cooperates. The family of companies span many industries and service firms. Examples: Mitsubishi, Mitsui, and Sumitomo.

In a recent book entitled *Winning Combinations,* J. Botkin and J. Matthews describe a coming wave of entrepreneurial partnerships between large and small companies. Large companies often have particular strengths in marketing, distributing, and selling as well as sources capital. Their weaknesses are apt to be innovation and speed in getting new products to market. The intuitive reaction of the manager of a large company, looking for specialized technologies that are not available in-house, would be to find a company in the marketplace that has what is needed, and to buy it. The problem with this approach is that once the acquisition has been digested and merged into the existing corporate structure, the innovative capacity of the target company will probably be gone. It will no longer generate the stream of innovations that the large company needed in the first place. It is like eating your seed corn. The alternative is to negotiate a strategic alliance instead. IBM has more than 500 partnerships with small companies and has acquired more than half a billion dollars worth of equity in these small firms. Asea, Brown Boveri in Europe owns or has equity investments in over a thousand small companies worldwide that provide power transmission and other electircal equipment.

The road to an enduring synergy between a large company and a small one can also be exactly the opposite: a spin-off, where a large company parcels out a portion of its operations and establishes it as a separate entity, with its own management and its own ownership (usually, the large company will retain at least a minority interest in the spin-off). Corning saw a new market for ceramic-based catalysts for power plants. Together with two Japanese partners, it started Cormetech, a separate corporate entity, to enter a mushrooming market created by federal environmental regulations.

On the surface, the rush for allies in an industry may look like chaos, but there is a clear pattern underneath it all. The companies are jockeying for positions that will enable them to have access to the advanced technology and to the marketing opportunities of the future.

Universities and Federal Laboratories: The New Breeding Grounds for High Tech Companies

The high tech economy needs to spawn a continuous stream of new technologies. Some of them are developed within the private sector itself. But increasingly, much new high tech will come from universities and federal laboratories. A key function in the future economy will therefore be the transfer of technology from the not-for-profit sector to private industry. The transfer of technology from the researcher to the businessman.

There is a need for new organizational formats that can move nascent technology from the laboratories of the academic sector to private enterprise. A biology researcher at a university may uncover new biotechnology that has immense commercial potential. That researcher may be a walking encyclopedia when it comes to specialized knowledge in molecular biology. But the odds are that he will do poorly as an entrepreneur.

The Japanese have an expression for that link between the laboratory and the subsequent successful commercialization: *kaizen*. *Kaizen* entrepreneurship recognizes that for every successful commercialization there are two innovators: one is the inventor of the new technique; the other is the manager who nurtures the fledgling company into a huge industry. The latter process may require a series of steps, each with a different innovator (or even a different firm). Americans often complain that the Japanese have "stolen" U.S. technology, such as color television, perfecting it, and then exporting it back to the U.S. But this is the kind of normal transition through stages that the Japanese expect from any original invention, including their own.

To smooth the transfer of technology, society needs new institutions and a new social framework. The precise nature and function of these institutions is not yet clear. No doubt the business incubator is part of the answer -- breeder institutions operated jointly by the private and the public sector and supplying a continuous flow of innovations and new products entering the marketplace.

The IC^2 Institute -- Innovation, Creativity, and Capital -- at the University of Texas is a novel type of academic institution, pioneering new ways of commercializing university science and technology. The institute has become a

The Vision of George Kozmetsky

George Kozmetsky belongs to that generation of researchers who in the early 1950s, armed with computers and a new kind of mathematics, set out to develop a theory of the scientific management of corporations. That new theory is called today operations research (OR) or management science. As a young man working at Hughes Aircraft, George Kozmetsky participated in the formation of TIMS -- The Institute of Management Sciences -- and later became one of the first presidents of the new organization. During the 40 years since, TIMS has grown to become a worldwide professional organization having thousands of members, working in such diverse fields as operations management, financial control, computerized inventory and cost systems, marketing science, logistics and management strategy.

There was a strong temptation to these young mathematicians and operations analysts to apply the new techniques to the business world. C.B. Thornton was rapidly building Litton Industries into the largest conglomerate in the world, and here George Kozmetsky got his first entrepreneurial experience. In 1960, together with Henry Singleton he founded Teledyne Inc. Using advanced operations research techniques the new corporation managed to land several large defense contracts. Eventually Teledyne was to grow into one of the nation's largest conglomerates, headquartered in Los Angeles.

Kozmetsky is the kind of person who believes that one person can change the world. He wanted to use the new techniques of scientific management to propel U.S. management and U.S. corporations to world-wide excellence. In 1966 a sleepy Southern state university approached him offering him the deanship of its business school -- the business school at the University of Texas at Austin. He had never lived in Texas before. It was a challenge that appealed to him. The rest is history. In 16 years he transformed that same business school into one of the finest business schools in the nation, ranked by U.S. News and World Report in 1987 as 11th nationally. The sprawling school complex now bears his name -- the George Kozmetsky Center for Business Education.

In 1977, George Kozmetsky and several university faculty colleagues founded the IC^2 Institute. The Institute quickly became established as an academic leader in the field of researching the spawning of new technology arising from space research and government defense programs, and the commercialization of biotechnology and medical engineering. Kozmetsky stepped down as dean of the business school in 1982 to work full time as director of the IC^2 Institute, while continuing to serve as executive associate for economic affairs at the University of Texas system.

In 1993, in recognition of his far-reaching influence in technology and education, Kozmetsky was awarded the National Medal of Technology by President Clinton. The medal was given in recognition of "his commercialization of various technologies through the establishment and development of over one hundred technology based companies that employ tens of thousands of people and export over one billion dollars worldwide." It is the nation's highest award for achievements in the commercialization of technology or the development of human resources that foster technology commercialization.

center of study of the creative and dynamic aspects of capitalism. It has a unique mission: to explore new organizational and management structures that can facilitate the commercialization of university-developed technology.

The institute is the brainchild of George Kozmetsky (see box). The philosophy behind the new creation is easy to explain: As the pace of commercialization of high technology accelerates, the nation's universities will have to assume a new role not only in the development of science and technology, but also in actually converting that technology into viable businesses. This requires not only new academic institutions (such as university-operated incubators and business parks), bridging the gap between the laboratory and the marketplace, but an entirely new mindset in the academic world. Historically, university people have tended to shun the idea of tainting their pure research efforts with such practical aims as making money. But times have changed. Universities need money, vast sums of money, to be able to hire top scholars and to buy expensive equipment. Many large universities, including the University of Texas, are these days heavily into the business of promoting and leasing university-developed technology to private business.

There is also another concern: the responsibility of the university for its students. Twenty years ago, the U.S. economy still generated large numbers of standardized academic job openings, like engineers or financial analysts. More recently, the market for university graduates has become splintered into niches, and students can no longer reckon that there is a job out there waiting for him or her upon graduation. Instead, many graduates will have to become self-employed. The university of the future needs to prepare its students for this possibility -- setting up new career paths across campus, opening up the possibility for young people to start their own businesses. Today, most universities have "placement officers" assisting their graduates. Tomorrow, universities will need "commercialization officers."

The Austin Technology Incubator was opened in 1989. It was funded by the city of Austin, the Greater Austin Chamber of Commerce, and Travis County; it is operated by the IC^2 Institute. Presently, it covers a suite of offices and a light manufacturing facility and wet lab, totalling 60,000 squarefeet. Twenty-seven high-tech startups currently reside in the incubator. The tenant companies are given three years to graduate and launch new businesses in biotechnology, telecommunications, computer, and software industries.

Since 1993, the IC^2 Institute operates two NASA technology commercialization centers: one at the Johnson Space Center in Houston, and another located in Silicon Valley. The Houston facility is tapping the Johnson Space Center's heavy support of medical research in space. For Ames, the focus is on computer hardware and software and materials. Both commercialization centers are modeled after the Austin Technology Incubator.

The NASA budget is tighter than ever. In the current political climate of deficit reduction, the Clinton administration is looking into means of reducing the cost of the space shuttle and other NASA programs. To prove its usefulness to the nation, NASA would like to intensify the pace of commercial applications of its technologies. Surprisingly few NASA technologies have succeeded the marketplace. The often quoted examples of teflon and Tang (the orange-flavored drink) actually existed long before NASA was formed. The cordless drill, developed by Black and Decker, is one of the few space technologies that made it. There is a potential for much more: growing crystals and manufacturing drugs in outer space, for instance. NASA officials are eager to licence space technology to the private sector. But NASA itself is poorly positioned to enter the commercial arena. New intermediaries, like the technology commercialization centers, and other know-how networks, are needed to bridge the gap.

References

For this chapter, see *The Economy as an Evolving Complex System*, eds. P.W. Anderson, K.J. Arrow, and D. Pines, Santa Fe Institute Studies in the Sciences of Complexity, vol.5, Addison-Wesley, Redwood City, CA 1988. A group of researchers at the Santa Fe Institute headed by Stuart Kauffman are currently working on economic models of "open technological web evolution." See also W. Brian Arthur, *Emergent Structures: A Newsletter of the Economic Research Program*, The Santa Fe Institute, Santa Fe.

"Peter Drucker calls the emerging form..." P.F. Drucker, "The Network Society", *The Wall Street Journal*, March 29, 1995.

"The Japanese have developed institutional formats of collaboration..." F. Phillips, "Technology Incubation, Japanese Style," *Technology Knowledge Activities*, Vol.1, No.1, 1993, pp.11-14.

Section entitled "What is a Strategic Alliance?" The IC^2 Institute publishes a journal entitled *Technology Knowledge Activities* featuring regular columns on "Leading-edge Technologies" and "Technology Alliances for Competitiveness." To get a copy of this publication, write the Institute, address 2815 San Gabriel, Austin, Texas 78705.

"In a recent book entitled 'Winning Combinations'... J. W. Botkin and J.B. Matthews, *Winning Combinations: The Coming Wave of Entrepreneurial Partnerships Between Large & Small Companies*, John Wiley and Sons, New York 1992. With a foreword by G. Kozmetsky, the IC^2 Institute, The University of Texas at Austin.

Box on George Kozmetsky: See also D.V. Gibson and E.M. Rogers, *R&D Collaboration on Trial*, Harvard Business School Press, Boston, Mass. 1994, pp 453-454.

CHAPTER 3

Messages, Images, and Robots

By communication one means the transmission of information (a "message" or an "image"). The transmission is from a sender, via a medium, to a receiver. There exists a market for information -- demand and supply for information. The demand is by consumers or corporations who want information, the supply by providers of messages and images -- media. There is a market price for information. There is a price for letters (postage), for telephone communication, for fax, for newspapers, for books, for private investment letters, for cable TV, for renting video, for movie shows, for admission to concerts and museums and theme parks.

The production of messages and images is a vast industry -- the information industry. Once the production chain involved just a single link like the runner from Marathon telling the Athenians about their victory over the Persians (a message), or veiled girls doing a belly dance in front of an Arab pasha in a desert tent (an image). With the advance of technology, the communication chains have become longer, involving entire networks of successive communication activities. To deliver a news segment on the evening news, hundreds of people may have cooperated, TV journalists filming and taping the news at the scene of the event, rushing the tape to the studio in a distant land, beaming it via satellite to the head office of the TV network, editing the news and transmitting it via TV stations or cable.

Innovations in the communications industry often means that existing media become obsolete. There are no semaphores any longer. (Semaphores were used by the French revolutionary armies in the late 1790's to relay information to Paris about their victories in the Savoy.) Even the telegram is about to become extinct -- if you send one these days chances are it will be phoned in to the addressee. The 78 rpm gramophone record has disappeared. There are no mimeograph duplicators in offices any longer. The mechanical telephone switchboards are collecting dust in junk yards.

The 16 mm home movie film that my father used in the 1930s remains today only as a medium for specialists. The 8 mm film that I myself started using as a young man was eventually replaced by the supereight. Now all home movie film is gone; instead there is the home video camera, both the beta system (developed by Sony but rapidly disappearing in many states) and VHS. In the early 1980s, RCA unsuccessfully tried to develop a video-disc type device.

Progress in the communications industry means several things: faster communications, communication over greater distances (in time and in space), more accurate or distortion-free transmission, the transmission of greater volumes of information, and the accessibility of the information transmitted.

Milestones of the Telephone	
1876	A.G. Bell demonstrates the telephone at the Centennial celebration in Philadelphia
1878	First commercial switchboard, in New Haven, Conn., serving 21 telephones
1893	Commercially wired broadcasting starts in Budapest (music, news, stock reports)
1921	First fully automatic switching system (in Omaha)
1926	Transatlantic telephony via submarine cable
1937	The coaxial cable
1946	High frequency microwave radio links
1962	Telstar satellite
1984	Divestiture of American Telephone and Telegraph
1984	Cellular phone introduced
1989	Integrated services digital networks (ISDN)
1995	The Federal Communications Commission auctions off licences for new high-frequency "personal communication services" (PCS)

Communication enables man to extend his eyes and ears and speech, to telescope his faculties as it were, to concatenate his own senses via media with those of other people, to see and hear and speak over great distances in space and in time. The result is to create larger aggregates of "grey cells" extending the confines of a single brain. In this sense, the advances in communication have led to better knowledge, understanding, and control.

Using the language of modern computer technology, it is possible to arrive at a common conceptual framework for all forms of communication. Consider the telegraph. Data are entered by the telegraph clerk tapping the Morse signals. The data are "processed" as they are converted into electric pulses. The receiver is a "peripheral."

The camera is processing data read by the optical eye (the lens). The data are "stored" in a memory device (the photographic film). The data are eventually "retrieved" through the development of the film. In television, the data are read by the TV camera and the microphone. In the case of live transmission, the data are thereupon immediately transmitted via wireless broadcasting or via cable to the TV receivers (peripherals). In recorded programs, the data are stored before transmission.

In a Rembrandt, the data (the object of the painting) were read by the eyes of the great master, processed in his creative mind, and stored on the canvas. An art museum is really a huge "data base." Using equipment of modern desk-top publishing, I can draw a color diagram on a computer screen with a light pen,

Milestones in Photography	
1839	The daguerreotype process (J. Daguerre)
1871	The dry plate
1888	Eastman Kodak box camera (George Eastman)
1900	Kodak's "Brownie," price $1
1924	35 mm film (the Leica camera)
1935	Kodachrome color film
1938-1948	Xerography (Chester Carlson)
1948	The Polaroid instant print camera (invented by E.H. Land)
1992	Eastman Kodak introduces its Photo CD system, a means for storing and editing images on compact disk

record it, and print it on a laser printer. I can store a bunch of such diagrams on a floppy disk.

The point that I want to make is that data processing has really existed conceptually long before the advent of the electronic computer. Quite generally, the communications industry involves the following steps: the recording of data, the processing and storage of data, the transmission of data to a close or distant receiver, and the read-out or print-out or display (or even speak-out, by a voice synthesizer) of the resulting messages or images.

A recent macro-innovation in the field of communications is digital communication, the conversion of sound waves and light waves to digital (binary) codes, the storing or transmission of such codes, and the conversion back to sound or pictures. Examples are the word processor, the CD disk and fiber-optic cables. In the word processor, letters are transformed into electric images, stored and retrieved. The CD disk (developed by Philips and Sony and introduced in 1983) can store more than 100,000 pages of text. A laser beam reads the information stored in the grooves of the disk. In fiber-optic cables, sound or pictures encoded in binary form are transmitted by a pulsed laser beam through a glass filament about as thick as a human hair. The contents of an entire CD disk can be transmitted in a few seconds. Digital communication gives an almost distortion free representation of the original message, and is much cheaper than the transmission of uncoded sound or light waves. In the future, most communication such as TV, radio, satellite communication, fax, cellular phone is going to be digital.

In 1988, AT&T together with European partners laid the first fiber-optic cable under the Atlantic Ocean, which can carry 40,000 simultaneous conversations. It can also carry data, fax and video. Another cable shared by U.S. Sprint and a British corporation is able to carry about 80,000 simultaneous telephone conversations. Two more fiber-optic cables were laid in the early 1990s.

Digitalization (or "digitization," the terminology is not yet settled) brings out not only the conceptual but also the technological affinity between conventional means of communication like the telephone and the gramophone, and the computer. The message that is transmitted over a fiber-optic telephone cable is not just "like" data. It is data. Digital data. The message that is read by the gramophone playing a Beethoven symphony recorded on a CD disk is data.

The Bewildering Future of the Telephone

When American Telephone & Telegraph Co. was deregulated and split up on January 1, 1984, the seven Baby Bells (Ameritech, Bell Atlantic, BellSouth, Nynex, Pacific Telesis, Soutwestern Bell and U.S. West) were each handed a local monopoly on their local lines. AT&T itself retained the long-distance part of the operations. The "consent decree" of the breakup bars the Baby Bells from manufacturing telephone equipment and from offering long-distance and information services including electronic publishing. But the Baby Bells do collect access fees when a long-distance call is routed through their local companies.

This sounds all fine in theory. But the real world is relentless product innovation and cut-throat competion. The competition comes from private long-distance carriers like MCI and Sprint. And, increasingly, it comes from cellular phone. Cellular systems transmit calls via analog radio waves beamed to local base stations clustered in "cells." The calls are relayed from cell to cell as a mobile caller travels. The base stations are wired to the regular phone network.

To stem the onslaught from the cellular communication companies, both the Baby Bells and AT&T itself have acquired cellular services of their own. BellSouth bought Mobile Communications Corporation of America, Southwestern Bell acquired Metromedia, and Bell Atlantic bought Metro Mobile Communications. And to top it all, AT&T bought a 33 percent stake in MCCaw Cellular Communications Corp in 1993. AT&T, said the aggressive chairman Robert Allen, aims at putting a cellular phone in everyone's hip pocket, and to provide the transmission for those phones over a nationwide "seamless" communications network. To many observers it seemed as if Humpty Dumpty was being put together again! More and more people will be making cellular local calls directly through AT&T, bypassing the regional companies.

The private long-distance carriers played the same game. In 1993, Sprint, the third largest long-distance company in the U.S. acquired the Centel Corp.

Fighting back, the Baby Bells are pushing to have the consent decree lifted. Several Baby Bells are striking up alliances with cable companies, preparing for the "electronic superhighway" of the future. Bell Atlantic and Nynex are pooling their cellular-phone operations to form the second largest company in the domestic wireless market, trailing only McCaw Cellular.

A quite troubling aspect of the esoteric services and fancy high tech solutions now being chased by the Baby Bells is that these companies are still local monopolies, with their rates set by regulators. The Baby Bells can offer all sorts of risky information services and video programming and equipment because they know that the regulators will always permit them a reasonable return on their *total operations*. The rate payers will be stuck with the bill.

In the meantime, the next step in the evolution of telecommunications technology is coming closer: personal communication services -- PCS networks. Subscribers will carry tiny cordless phones with computing, faxing and video features. These Star Trek-like communicators can be tucked into a coat pocket or into a small purse. PCS will be fully digital. The Federal Communications Commission began auctioning the station licenses in 1995 (collecting some windfall government revenue). PCS are transmitted at much higher frequencies than regular radio signals. Since their broadcasting range is shorter, they require more cells and more base stations.

The idea behind the PCS is revolutionary: in the future a person will have one telephone and one telephone number that will follow him or her from home to car to office to city throughout the world.

Supercomputers and Parallel Processing

The first all-electronic computer was the ENIAC (electronic numerical integrator and computer). It was a 30-ton machine developed by J.P. Eckert and J.W. Mauchly and built at the University of Pennsylvania in December 1945. It had 18,000 vacuum tubes. It was used by the Army to calculate ballistics tables for new weapons. Soldiers stood by with baskets of replacement tubes.

The transistor was developed by three scientists at Bell laboratories, W. Shockley, W. Brattain and J. Bardeen in 1947. The transistor is a midget device which turns current off and on as it moves over a controlled path within a solid block of semiconductor material. The "second" computer generation began in 1959 when machines employing transistors became commercially available.

The "third generation" came with the integrated circuits. During the late 1960s and the 1970s electronic components were further miniaturized. In a microprocessor all the logic and control circuitry that is necessary for interpretation and execution of instructions is etched on a silicon chip. In semiconductor memory, each memory cell is similarly etched on a chip.

Modern computer chips pack hundreds of thousands, or millions, of transistors on one piece of silicon. Such chips are capable of handling millions of instructions per second (MIPS). New techniques of making chips are being presented almost every week. Recent advances include DRAM (the acronym is pronounced "dee-ram" by the cognoscenti and stands for dynamic random access memory) and RISC (reduced instruction set computing). Future "superchips" will contain up to one billion or more transistors on a single chip.

Supercomputers are the most powerful general purpose computers available at any given point in time. In 1977 Cray Research emerged as the preeminent supercomputer manufacturer. The Cray-1 system was essentially without head-to-head competition. In 1985 Cray stock had grown in value by a factor of over 100 since its initial public offering. This was the highest ten-year growth average of any publicly traded stock. An international scramble ensued to enter this highly lucrative market. Current manufacturers include Cray Research, Cray Computer (a second supercomputer company started by Seymour Cray in 1989), ETA Systems, and Amdahl in the U.S., and Fujitsu and Hitachi in Japan. Supercomputers allow engineers to test new aircraft designs using computer models instead of costly trials of physical models in wind tunnels. In the automobile industry, supercomputers are used for design and crash-testing purposes supplementing costly physical testing. Financial firms, including Goldman Sachs and Nomura Securities, run large-scale financial models of their securities and loans portfolios on supercomputers.

In parallel processing, a software program is broken into different tasks that are distributed among more than one processor. Multiple instruction streams are able to perform operations on multiple data streams concurrently. The BBN Butterfly Plus computer is a 14 processor machine, each processor having some memory of its own, and accessing the memories of the other processors through a network. Thinking Machines Inc. of Cambridge, Mass. manufactures a 64,000 processor computer, dubbed the Connection Machine. It is believed that through such "massively parallel" computing it shall be possible to simulate better the information processing in the human brain. Thinking Machines Inc. has sold its computers to Dow Jones & Company for a new electronic information retrieval service called Dow Quest. The service lets an unskilled user browse rapidly through a vast library of information from business newspapers and magazines. As parallel machines become more widely available, users realize that parallelism comes natural to many applications. Tracking the flow of aircraft traffic on the high-altitude routes can be allocated to multiple processors, one for each route.

Parallel processing requires not only new computer hardware but also a new way of writing software. Actually, it requires a new kind of mathematics. A given problem is broken into many parts. Each problem part is pursued independently for a while, the intermediate results are reported back to a solution center, and remaining matters to be cleared up are again broken into many parts and delegated for individual analysis and collation. Such parallel algorithms are now being developed for many standard problems in operations research, and mathematical management control systems.

Artificial Intelligence

The term "artificial intelligence" (AI) has to be understood and used with some caution. No existing computer is intelligent in the sense that it can figure out things on its own, beyond what it has been programmed to do.

Computers can play chess. It is easy to program a computer so that it knows the permitted moves of all pieces. The computer will analyze possible scenarios for the next few moves. This analysis will include both an evaluation of the possibilities of the own player, and the possible counter moves by the opponent. The computer may consult a library of standard opening moves. There are even tournaments for chess-playing computers.

The most successful chess-playing computer to date is Deep Thought, built by Taiwan-born F. Hsu at Carnegie Mellon University in Pittsburgh. It is able to examine up to a million chess positions per second. Deep Thought has whipped

Milestones in Television	
1932	RCA (Radio Corporation of America) under the leadership of David Sarnoff demonstrates all-electronic television (with a cathode-ray tube for the receiver)
1935	Regular broadcasting service in Germany, 180 lines per picture
1936	BBC starts broadcasting in London, based on the EMI system (Electric and Musical Industries), 405 lines per picture
1941	Regular TV broadcasting in the U.S., 525 lines
1954	Public broadcasting in color in the U.S.
1956	Videotaping of TV programs developed by Ampex Corp. (early devices were the size of a washer-dryer);
1988	2 billion viewers worldwide witness the opening ceremonies of the Seoul Olympiad
1988	Standards for HDTV (high definition TV) set by the National Television Systems Committee, 1125 lines

a number of grandmasters. In October 1989, in a hall in the New York Academy of Art, the current grandmaster Garry Kasparov met Deep Thought. The first game was a Sicilian Defense. After eight moves, Deep Thought started to slip. At move 13 it failed to castle (a standard measure to safeguard the king), and by move 21 it fell into a trap. Kasparov exchanged his dark-squared bishop for one of the computer's knights. A bishop is usually considered to be worth more than a knight, but in this particular position the exchange gave Kasparov a clear strategic advantage. Kasparov deftly maneuvered his knight into a controlling position, and the computer resigned. The second game was a Queen's Gambit, but Kasparov deviated from the well-known sequence of moves. The computer blundered and lost again. As the outcome of the match demonstrated, cunning and boldness can still outwit the sheer numbercrunching ability of a computer.

What people usually have in mind when they talk about artificial intelligence is a "high level" software language. The lowest of all languages is machine language. Next comes "assembly" -- packages of machine language instructions. Then there are the standard languages like BASIC where the commands refer to individual logical steps like reading, adding, subtracting, looping, writing. High level languages are problem oriented rather than procedure oriented. Spread-sheet programs are high level. Computerassisted design (CAD-CAM) is high level.

Three forms of artificial intelligence need special mention: expert systems, pattern recognition, and natural-language comprehension. An expert system is a knowledge-based software system that provides a computer with the capacity to make decisions for solving complex problems. A physician enters medical data onto a computer screen, the expert system consults a library of all known diseases and medical afflictions, and suggests a diagnosis. A lawyer enters the pertinent facts of a legal case, the expert system consults a legal library, and lists similar earlier cases and their outcomes. A travel agent enters a desired travel route, the expert system consults a library of all air connections, suggests airlines and itinerary, and figures out the price of the ticket.

Pattern recognition involves the decoding of messages in the presence of noise and distortion. The computer reads a message, converts it into digital form and compares it with patterns stored in the computer's memory. The computer isolates relevant features, and filters out unwanted disturbances. There are scanners that can read typed text and translate it to electronic images, using it as input to word processors or voice synthesizers. The next step will be pen-based hand-held peripherals that recognize cursive handwriting (like Apple Computer's Newton MessagePad).

A number of systems are already on the market that can recognize a limited vocabulary of spoken commands. Microsoft, the software giant, has introduced a Windows sound system that permits the user to control basic functions of popular software (such as saving a file) by speaking into a microphone. A commercial air traffic control simulator permits the trainee to issue commands like "American 931, climb and maintain 8000" or "Cherokee 47N, turn left heading 250." Several telephone companies offer voice-activated services. When the caller recites an identification number, the system calls up the user's voice print to confirm the person's identity. Once identified, the caller may engage in a limited conversation with the system, asking for billing information or to be connected with persons by their names rather than telephone numbers .

The computer uses a process called "dynamic time warping" to identify voice commands. It breaks up a sound into slices of time each only a few milliseconds in length. It analyzes the energy and frequencies in each time slice, matching them against a master data base. In a "speaker-independent" voice recognition system, that data base has been developed from thousands of different voices. In a "speaker-trained" system, it has been assembled during a prior training session. Having established a match, the computer then acts on the command.

A natural-language software system will query a data base on a specific subject. Such a system needs a vocabulary and information about elementary grammatical rules and common violation of those rules. For instance, a computer program developed at the University of Texas enables the user to enter simple mechanics problems (such as finding the point of gravity of a ladder leaning against a wall) in plain English, the software will thereupon interpret the problem formulation, draw a picture of the objects and their location, and solve the problem. And computational linguists are working on automated translation systems, translating English into German or French. To do this, the computer needs to be able to tear sentences apart word by word, identifying subjects, objects, verbs, and clauses. It needs to be able to "understand" each sentence -- that is, match it against recognized grammatical principles.

Much human effort has been spent on designing electronic computers that are able to learn. A computer learns if it is able to adjust its own software as it gains experience from various tasks. The computer compares its own performance with some desired norm of performance -- a criterion of "better" performance -- and adjusts its software accordingly. The logical feasibility of such systems was recognized early and was explored mathematically by the U.S. mathematician N. Wiener (1894-1964). There are computers that learn in a primitive way, for instance a mechanical mouse that runs through a maze and learns to make the right turns to reach a desired destination. More advanced learning systems are based on stochastic decision rules, where the a priori probabilities of each possible course of action are adjusted in light of previous experience. Consider for instance a computer playing some simple game involving a loaded dice. At the outset, the computer will assume that all six outcomes of a throw of dice are equally probable and equal to 1/6. But, as the game is being repeated, the computer may be programmed to examine the relative distribution of the various outcomes. If the outcome of "3," say, tends to be more frequent than 1/6, and the outcome "4" less than 1/6, the computer may be instructed to adjust the theoretical probabilities. Such action will improve its net return made from the game, and the computer will continue to adjust the probabilities until the return is maximized. The computer has "discovered" the loading of the dice, and has "learnt" how to adjust its play in response to this insight.

Many researchers have argued that it would be necessary to build an entirely new kind of computer, a "neural" computer, to mimic the actions of the human brain, including perception, understanding, and learning. While attempts to build such computers have been made, the real breakthrough has been in the construction of new software called "neural networks" which can be run on standard digital computers. Neural networks learn. For instance, a neural network can be set up to search for faint information signals present in noisy data like stock market data. To accomplish the search, the network is first "trained" using historical data. Once the system has learned from these training sessions, it is ready to go, analyzing current data.

In principle it would seem to be possible to develop a natural-language software system that learns just the way a child learns a language, gradually expanding its grammar and vocabulary in the light of experience. But such systems still belong to the future.

Early mathematicians were stunned by the miracles of electronic computers. They approached with awe concepts that we today take as the most natural thing in the world. John von Neumann, considered the father of the modern computer, entitled his pathbreaking treatise on the subject "On the Theory of Self-Reproducing Automata." A software program can self-reproduce itself because it can include instructions to duplicate itself. That is really no big deal. Any kid these days knows what it means to copy a cassette tape, or to duplicate a video tape. But in the early 1940s these thoughts were almost heresy. Living organisms can self-reproduce. To suggest that creations of man can self-reproduce seemed to challenge the Creator Himself.

Today, the common attitude is quite the opposite. It is taken for granted that computers will eventually be endowed with most of the faculties possessed by humans. If our brain simply is a "computer made of meat," perhaps the computer engineers of the future shall be able to construct computers that can feel pleasure and pain, and appreciate beauty and humor. In brief, why should not the computer of the future be endowed with a mind? Roger Penrose, English mathematician and cosmologist and working with Stephen Hawking on black holes and the origins of matter, has protested strenuously against such inference. In a recent book entitled "The Emperor's New Mind" he suggests that the mind of the computer is -- and will always remain -- nonexistent, just as the emperor in Hans Andersen's well known fable actually was naked, without clothes. To Penrose, there is a vast unknown in our physical understanding of the working of the human brain. Questions of philosophy and psychology have their role to play when it comes to attempting to understand the nature and function of consciousness. Like Newton and Einstein, Penrose has a profound sense of humility toward the physical world. Mathematicians are occasionally allowed to glimpse part of a page of "God's book" in which all the best proofs are recorded. For a fleeting moment mind makes contact with objective truth. But only for a moment.

Robots: Slaves or Masters?

Up to this point, I have dealt only with information systems that generate output in the form of messages or images. Now, I am set to deal with outputs that take the form of some kind of action of a "machine." I shall call a machine controlled by an information system a robot.

The word "robot" is of recent origin and was created by the Czech author K. Capek. It means "forced labor." But the idea of an omnipotent slave who would obey the bidding of his master is as old as civilization itself. In Jewish folklore, "golem" was a mythical figure, part slave and part devil (see Psalms 136, 39). The genie in the bottle played a similar role in the Arab world. In Goethe's tale about the sorcerer's apprentice, the apprentice conjures up sinister slaves whom he cannot control. (The reader may remember Disney's animation of this story in *Fantasia*, to the music of Paul Dukas.)

It is helpful to distinguish two categories of robots: those that form a closed and repetitive system, and those that allow for interaction with their environment. The pendulum clock and the spring-driven clock belong to the first category. So does a mechanically animated bird, a washing machine, and a toaster. The second category involves feedback or control.

An automatic elevator, a vending machine, or a microoven constitute an intermediate category. There are several repetitive systems to choose from, and the customer decides by a simple command which of them shall be activated (pressing the button or setting the controls). Such "menu"-driven robots can actually be very sophisticated. Consider the case of an automated warehouse, where a customer enters the identification number of some spare part on a computer screen. The information system consults its memory and identifies the physical location of the box of parts in the warehouse. It causes an unmanned forktruck to move down the aisles of the warehouse until it reaches the correct aisle, to turn into the aisle, to approach the location where the box is stored, to telescope the forklift to the appropriate height, and to pick the desired part with a mechanical hand. The forklift thereupon reverses its motions, returning to the front desk with the desired item.

In the Jacquard loom (using a punched paper tape to define the desired pattern of the woven fabric) the menu of commands is practically unlimited. There are as many different punched paper tapes as the fabric designer may care to dream up. Similarly, in an industrial robot in the automotive industry spraying paint or performing precision welding, the movements of the robot may be reprogrammed when called for, guiding the robot arm manually through a desired sequence of movements. In some advanced industrial robot systems the programming of the robot is effected from an entire database that may be updated periodically. Examples are robots for aircraft precision machining, and propeller grinding guided by a laser measurement system.

In feedback, a robot compares the actual performance of a physical system with stored information about desired performance. A deviation will trigger action to correct actual performance and bring it into line with desired performance. In an ordinary room thermostat the desired performance is indicated by the temperature setting. If the actual room temperature falls below this level, the thermostat perceives such deviation, and takes required action. Another simple example is the automatic door opener.

Software systems directing robots and involving at least some degree of feedback are getting extremely common in the last decades of the twentieth century. The electronic cash teller machine is such a system. You enter a password. The computer compares the password with its list of acceptable passwords and clears you for a transaction (feedback). You issue an instruction to withdraw cash from your checking account. The computer compares the desired withdrawal with your current balance and pays the money (feedback). The computer updates the account.

An automated parking garage is another such system. (For some reason, most parking garages in the U.S. are still manned. Travelling in Scandinavia recently, I found automated garages in supermarkets, hospitals and in town squares.) As you enter the garage, a meter spits out a ticket. You take the ticket and the entry gate is raised. You park your car and validate the ticket, inserting ticket and money into an electronic validator. Having done your errands, you return to the car, approach the exit of the garage and insert the validated ticket into the reader. The system computes the parking time and parking cost and compares it with the amount of money already paid. It raises the exit lever, or instructs you to return to the validator (feedback).

A very similar software system is used in automated libraries. When you leave, an electronic detector checks that the books you carry under your arm are "demagnetized." If not, an alarm will sound (feedback). Or, consider the plethora of electronic home security systems, reading the inputs delivered by movement detectors, sound detectors, smoke detectors and what have you, and being ready to dispatch armed guards, the police, the fire brigade or the EMS (feedback). Australians have developed a sheep shearing robot which shears sheep automatically under real-time control. Feedback, from perception to action, takes time. In the meanwhile, the actual performance being monitored may have changed. The target with which the performance is compared, may also be moving. If great speeds are involved, it may be crucial to anticipate correctly such movement. Consider the case of an automatic guidance system for an air-to-air missile (fired from one airplane toward another). The target is detected by the host aircraft radar. The missile is equipped with a self-contained radar which is locked onto the target. The missile is launched. The missile guidance and control system continuously computes the relative path of the enemy airplane and extrapolates (= makes a statistical forecast of) where the target is going to be located in the next few seconds. It adjusts the guidance controls of the missile based on this forecast.

A Visit To Delbert Tesar's Laboratory

Delbert Tesar, a professor of mechanical engineering at the University of Texas at Austin, conducts one of the largest robotics research groups in the country. Previously, he was the founder and director of the Center for Intelligent Machines and Robotics at the University of Florida.

In his laboratory on the university campus, Tesar is developing a system of generic parts -- shoulders, elbows, wrists, and knuckles for robot arms. They will be assembled into a toolbox of robot parts run by special motors and commanded by microchips. According to Tesar, robotics requires a marriage between mechanical engineering and electronics.

In a corner of the lab sits a 2000 pound industrial robot, a state-of-the-art assembly-line industrial robot. It has a strength-to-weight ratio of about 1 to 100. The two fingers on the robot arm are quite weak in relation to the great weight of the robot. It makes precision work difficult even with advanced electronic control. Tesar is working on lightweight robots that would operate with a strength-to-weight ratio of 1 to 1.

Tesar envisages the future robot as a fully integrated and self-contained generic machine system capable of performing a wide spectrum of precision operations completely programmable by the designer of the product. He ticks off the following tasks to be performed by robots in the future: precision light machining, micro-manipulation, micro-surgery, human augmentation for the handicapped, prospecting and mining on the ocean floor, space station assembly and repair, battlefield operations, rapid runway repair even under attack.

The Department of Energy is funding the development of a lightweight robot to make repairs inside a nuclear reactor. Tesar's group of 38 graduate students is designing the software to be used to command the robot.

Today, almost all robots are designed one at a time at exceptionally high costs of resources and time. Tesar sees a pressing need to achieve an architecture and a standard of technology for robotics in the same sense that the IBM computer has become a standard for PC clones.

In the case of a cruise missile, the map of the territory to be overflown is digitized and stored in the missile memory. The radar on board the missile compares checkpoints on the map with actual observations and a robot mechanism adjusts attitude and speed to maintain the cruise missile on its desired approach path. If the missile detects that an enemy missile has locked on it, it takes evasive action (diving away).

In the war with Iraq, the U.S. fired Tomahawk cruise missiles toward strategic targets in Baghdad. As the Tomahawk nears its target, it takes visual readings, illuminating the target with strobe lights at night. The pictures are compared to data stored in the computer so the missile can make last-minute adjustments in its course. The Tomahawks were able to locate specific buildings in the city, destroying electronic command centers but leaving other parts of the buildings intact. The missiles were launched from attack submarines positioned in the Red Sea. The prime contractors of the Tomahawk were at the time General Dynamics Corp. and McDonnell Douglas Corp.

Spacecraft are also navigated by feedback. A celestial navigator calculates the position of the craft from star patterns. An inertial navigator receives responses from a gyroscope. Information from either navigator is processed by an on-board computer, and is used to control the guidance rockets.

Satellites and space junk are rapidly filling outer space. There are well over 5,000 orbiting objects larger than a baseball already under surveillance by the North American Air Defense Command. Many of them are broken satellites or satellite parts that should be repaired or retrieved. The space shuttle Discovery brought back for repair two malfunctioning satellites -- Palapa B2 and Westar 6. In the future, a robot called the Flight Telerobotic Servicer will repair or refuel satellites in space. It will also be used to retrieve tools or even astronauts that have accidentally been separated from the space station (if it is ever built). Approaching the object, the robot uses three-dimensional imaging, robotic arms, and gripping devices to capture it.

Dante, a eight-legged robot built at Carnegie Mellon University, was sent into the smoldering mouth of Mt. Erebus, an Antarctic volcano. It crawled down the crater wall of the active volcano, reached the lava lake on the crater floor, retrieved minerals and conducted a variety of experiments.

The mathematical theory of control and feedback was developed by the U.S. mathematician N. Wiener. He called the theory "cybernetics" (the word being derived from the same root as the word "govern"). Wiener was also keenly interested in the effects of computers on society, and in philosophical issues such as the relations between creativity and machines, or, to use his words: "God" and "Golem." His book *God and Golem, Inc.* is about a possible future of cooperation between man and machine.

In more modern terms, what Wiener brought up was the scope of interactive systems, of a dialogue between man and machine. Since I believe that the concept is important, let me put a label on it -- IMMS -- interactive man machine systems. The future of communication systems is not some hideous omniscient monster that tramples man under his feet. Quite on the contrary, IMMS is going to enable us to release the God-given creative powers in us and to devote ourselves to "the human use of human beings" (another catching phrase of Wiener's) rather than the mechanics of information gathering, information processing, mechanical action, and mechanical control.

IMMS is probably set to blossom around the middle of next century. These systems are still in their infancy. We can see the beginning of what is going to come in the modern automobile, with computers checking and controlling the injection system in the engine and the disk brakes. That is, a human person is still very much in control, with a multitude of computers and robots responding to the touch of the foot upon the accelerator or the brake. An other early example of IMMS is the automated editing and printing of a modern newspaper. Journalists around the world submit their stories digitally, via word processors and data links, to the editing office. The editor reads the stories as they arrive flashing in front of his or her eyes on a data screen. The editor assigns priorities to the news bulletins. The computer prepares tentative layouts of all pages, based on these priorities and on preset rules. The editor checks the proposed layout on a screen and makes changes. The computer directs the printing presses in several distant cities.

A U.S. firm offers computer-aided design of women's swimsuits. The swimsuit is created electronically on the computer screen, and the designer and the customer (a representative of a chain of department stores) discuss changes together. Once agreement has been reached, the swimsuit goes into automatic production.

A rather surprising example of such symbiosis between humans and robots is the "power glove" and "data suit" now being developed by NASA and to be worn by the spacestation servicing robot that I have already mentioned. A human operator onboard the space station, or even safely ensconced in a ground control center, mimics the required finger, arm, and body movements. Sensors pick up the movements of the operator and translate them into corresponding telerobotic actions in outer space.

In the next century, with computers and robots extending the faculties of their operators beyond present comprehension, our grandchildren will know more, see more, and reach out further than what is possible for the cleverest among us today.

Slave or master? The robot is about to transform us earthlings into cosmic rulers. Call it what you will.

Bibliographic Notes

To the economist, the subject matter of information, communication, and control is basic to the understanding of the evolving economy. For one thing, as developed in A.L. Norman, *Informational Society: An Economic Theory of Discovery, Invention, and Innovation*, Kluwer Academic Publishers, Boston 1993, it traces the chain of inception, development, and dissemination of new technology. At any given point in time, to a single corporation or to an entire industry, there exists a pool of available information, and a potential for acquiring additional information. At each step, the process of acquisition of new technological know how is at each step guided by economic considerations. New know how can always be acquired, at a cost.

But the economics of information goes even further. It also encompasses the acquisition of entrepreneurial competence and experience, thus setting the parameters for the destiny of corporations and of industries.

Rapid innovation in the information and communications industry affects every aspect of economic life, from manufacturing to distribution and marketing. In other words, information is economic change. And so, the economics of information is the economics of change.

All these subjects of discussion will duly be explored in subsequent chapters. As a starting-point, I have chosen to give a brief account of the hardwares and the softwares of the current information explosion. My approach is that of a network, and the dissemination and processing of information along a communication network. The links of those networks can be fiber optic cable, radio beams, satellites, and data networks. The messages passed along the networks can be information, or instructions to relays or machinery ("robots") to perform pre-programmed tasks.

References

Boxes on the telephone and photography. I have gleaned the information from many sources, including F. Williams, *The New Telecommunications, Infrastructure for the Information Age*, The Free Press, New York, 1991.

"In 1988, AT&T together with European parters laid the first fiber-optic cable..." AT&T national advertising, Dec. 14, 1988.

"Supercomputers are the most powerful ..." See ref. [4]. Also, L.M. Thorndyke, "Supercomputer Systems Markets," in ref. [6].

"Parallel processing requires not only new computer hardware but also a new way of writing software..." One of the first publications in this rapidly expanding field is D.L. Miller, J.F. Pekny and G.L. Thompson, "Solution of Large Dense Transportation Problems Using a Parallel Primal Algorithm," *Management Science Research Report* No. 546, Graduate School of Industrial Administration, Carnegie-Mellon University, June 1988.

"Milestones in Television." D. Sarnoff, *Pioneering in Television: Prophecy and Fulfillment*, RCA Technical Institutes Press 1946 and Looking Ahead, Mc Graw Hill, New York 1962. I have also consulted C. Carbonara, "HDTV," in *The United States and Japan: Shared Progress in Technology Management*, ed. by S. El-Badry, H. Lopez-Cepero, and F. Y. Phillips, IC^2 Institute, University of Texas at Austin, 1993.

"The most successful chess-playing machine..." J.A. Tannenbaum, "Champ to Chess Computer: Come Back in '92," *Wall Street Journal*, Oct.24, 1989. -- Matched in London in 1994 against a Pentium-based computer named Genius 2, Kasparov lost.

"The manufacturer of the system even offers a computer game version ..." The manufacturer is Wesson International, Texas. Another successful flight-simulation game is Microsoft Corp.'s Flight Simulator.

"Roger Penrose, English mathematician and cosmologist..." R. Penrose, *The Emperor's New Mind. Concerning Computers, Minds, and the Laws of Physics*, Oxford University Press, New York 1988, and, by the same author, *Shadows of the Mind: A Search for the Missing Science of Consciousness*, Oxford 1994.

"The Australians have developed a sheep shearing robot..." D. Tesar, "An Outline of the Problem and Proposed Solution to Our Weakening Economic Condition in Civil Sector Manufacturing," in *The U.S. Competitiveness Challenge: Background Papers*. Compiled by the IC^2 Institute, The University of Texas at Austin, January 1988.

Box on Delbert Tesar. I have used the following documents: D. Tesar, *A Proposal for a Program in Mechanical Engineering for Precision Machines in Manufacturing (Exexutive Summary)*, Department of Mechanical Engineering, The University of Texas at Austin, October 12, 1990 and D. Tesar, "Thirty-year Forecast: The Concept of a Fifth Generation of Robotics- -The Super Robot," in *The United States and*

Japan: Shared Progress in Technology Management, eds. S. El-Badry, H. Lopez-Cepero, F.Y. Phillips, The IC^2 Institute, The University of Texas at Austin, Austin, 1993.

"Consider the case of an automatic guidance system for an air-to-air missile... " I would like to thank Richard Drury, retired general in the U.S. Air Force, for helpful discussions.

"In the future, a robot called the Flight Telerobotic Servicer..." *Aviation Week and Space Technology*, June 11, 1990, p. 28.

"Dante, an eight-legged robot ..." *Carnegie Mellon Magazine*, Vol. 13, No. 2, winter 1994, pp. 10-12.

"The mathematical theory of control and feedback was developed by the U.S. mathematician N. Wiener..." N. Wiener, *Cybernetics or Control and Communication in the Animal and the Machine*, the M.I.T. press, Cambridge, Mass., 1948, 2nd ed. 1961. The following two works are non-technical: N. Wiener, *God and Golem Inc.*, The M.I.T. press, Cambridge, Mass. 1964 and *The Human Use of Human Beings*, Houghton Mifflin Co., Boston 1950, rewritten edition Doubleday Anchor Books, Garden City, New York 1954.

CHAPTER 4

The Technology Frontier

To economists, the word production means the generation of goods and services that have a value in the marketplace. Thus, production is not just the manufacture of an automobile or a synthetic drug, it is also computer consulting and open heart surgery.

A technology is a particular way of organizing production, converting inputs like skilled and unskilled labor, machinery and equipment into desired outputs. Examples of technologies are offset printing, growing tomatoes in industrial greenhouses, credit cards (the output is financial services), the iron lung.

The technology frontier separates known and economically viable technologies from those that are still under experimentation or have not yet been demonstrated to be commercially feasible. The frontier technologies are those at the "leading edge" or the "cutting edge" of present knowledge. They are efficient. Back, behind the frontier, are the outmoded or inept or inefficient ones, those that represent suboptimal technical practice or just bad management.

The technology frontier draws a line of demarcation between the known and the unknown, between the past and the future. It is the borderland of our present material culture. The frontier is being pushed forwards through innovations and through product development. Innovations are major discrete jumps in technology, like fiber optics, communication satellites, parallel computing. The term "macro-innovation" is sometimes used to denote the emergence of an entire new family of technologies like electricity, the combustion engine, the automobile, the airplane, computers. Inside each such macro family, there occur innovations. For instance, the macro-innovation of the airplane and commercial airlines was followed by innovations like the all-metal airplane, transatlantic flight (Charles Lindbergh), the autopilot, cabin pressurization, the jet engine, supersonic travel (the Concorde). The technology frontier is also being pushed forwards by a continuous sequence of minor changes in product design.

The Rise and The Fall of the Passenger Train

It is a pity that World Exhibitions are destroyed. Imagine that we could enter today the first exhibition in London in 1851 joining Queen Victoria and her consort Prince Albert in the opening ceremonies. The famed Crystal Palace had about 1 million square feet of glass. At the main entrance were two imposing representations, one of Courage (a statue of Richard Coeur de Lion) and the other of Power (a solid 24-ton block of coal). This was the age of coal. There were exhibits of art and handicrafts. And there was Machinery, like the first mechanized farm implements.

Or, imagine that we four years later, in 1855, could have joined Napoleon III and Eugenie in the opening of the Exhibition in Paris in the great Palais de l'Industrie. The exhibits were arranged in concentric circles with art in the center, and technology at the periphery. The building itself is still left in Paris and houses a museum. The World Exhibition in St. Louis in 1904 was opened as President Theodore Roosevelt pressed a button in the White House which by way of long distance telegraphic cable initiated the electrically illuminated water fountains. Today, we do not need to wait four years for the next world exhibition: the tourist can instead travel to Disney's EPCOT Center ("Experimental Prototype Community of Tomorrow") at Orlando to find anything that a world exhibition would display.

The reader may have visited a technical museum with a display of a series of vintages of the ordinary telephone, starting with Bell's original prototype, passing through old wall models with a swivel, models from the late 1920s made in the plastic material Bakelite, eventually reaching our own age with push-buttons. And one could add to that series the cordless phone, the TV phone (enabling the callers to see each other), and cellular phone. For most modern products one can trace such a chain of evolution.

The passenger train was one of the glorious products of the machine age. The Pennsylvania Railroad had operated several famous daily trails daily between St. Louis and New York. One of them was the eastbound *New Yorker*, another the westbound *St. Louisan*. In 1927 they were rechristened the *Spirit of St. Louis* in commemoration of Charles A. Lindbergh's transatlantic flight in that famous plane. The finest St. Louis to New York train of the "Pennsy" was the extra-fast and extra-fare *American*. Between the 1890s and 1950, the scheduled time from St. Louis to New York was reduced from 28 1/2 hours to 20 hours. Steam engines, which propelled all trains until the mid-1930s, were gradually supplanted by sleek electro-diesel locomotives for freight and passenger use. Streamlined lightweight trains designed to reduce wind resistance were introduced. Air-conditioning made its appearance on Pullman cars. Separate roomettes and bedrooms afforded greater space than sleeping car berths. The *American* featured an office secretary- stenographer, barber, valet, and telephone service from the train at its terminals. There were books, magazines and newspapers in the lounge-observation car. Women patrons had a manicurist at their disposal. Dining was advertised as an "epicurean masterpiece." In its heyday, the Union Station in St. Louis had the greatest number of tracks of any railroad station in the world. Twenty-two large railway corporations operated their trains out of the station.

The evolution of technology occurs in waves: a new superior technology replacing an earlier one. Each single technology and each single product design goes through a life cycle of commercialization, adoption, growth, and eventual decay. Why are there such life cycles? In 1786 Edward Gibbon published the first volume of his *The Decline and Fall of the Roman Empire*. Gibbon was steeped in the eighteenth-century belief in strict causality and extended it to the sphere of history. He did not just describe historical events. He wanted to explain them, in terms of causes and effects. Other historians were to follow in his footsteps (Spengler, Toynbee). Is it, similarly, possible to develop a causal hypothesis of the life cycle of a modern product? The rise and the fall of the passenger train? The rise and the fall of the vacuum tube? The rise and the fall of computer consulting services?

At least we can make some preliminary observations. During the upswing of the life cycle, the producer gains experience of the manufacturing and distribution processes. Marginal costs come down. This is the so-called "learning curve." Learning curves have been documented in many organizations, in both the manufacturing and service sectors. During the first year of the production of Liberty ships during World War II, the number of labor hours required to build a ship fell by close to a half, and the average time it took to build it decreased by 75%. Learning curves have also been found in kibbutz farming, nuclear plant operating reliability, and even success rates of new surgical procedures. It is a pervasive phenomenon. Sometimes it is referred to as a progress curve, an experience curve, or learning by doing.

At the same time, the new and superior demand pattern is diffused among the purchasing public. The ranks of the first "innovating" buyers who accept the new product on their own are joined by "imitating" buyers, that is, people who are influenced in their new product purchasing decisions by other people. The purchases of both categories swell under the barrage of advertising efforts, promotion, and a falling price. Mature products representing state-of-the-art technology define the technology frontier.

The downswing of the life cycle sets in when the corporation is not able to keep pace with the creativity and the change of its competitors. The consumers, restlessly looking for superior consumption alternatives, turn to more advanced product designs. Brand loyalty is a fickle thing. As the buyers desert the product and sales dwindle, the life cycle races downhill. The more sophisticated the technology, the shorter the life cycle.

To stay on top of these complex dynamic processes, a new kind of management is needed: product cycle management. A corporation needs to bring on line a continuous stream of new product designs, each in some sense "better" than the preceding, just in order to survive. The corporation therefore must have in the pipeline a larger number of new products than those it will actually end up with. It must see to it that there is a sensible timing of the new product designs, so that an optimal time flow of new designs hits the market. There must be a second generation and a third generation of the current models under development. Some of these will fail; a few will survive. Those that do survive have only a limited life span, and will soon need to be replaced by superior models. The scheduling problem consists of determining when each vintage should hit the market (commercialization) and when it should be withdrawn (technological or economic obsolescence). At any given point in time, a corporation will manufacture and market an entire line of products. Some of the individual products are recent market introductions, others are approaching the end of their economic life span.

What is at issue is really the management creating its own future. And this future must change at an ever increasing speed.

An early master -- and an evil master -- of product cycle management was Friedrich Alfred Krupp (1854- 1902), grandson of founder Alfred Krupp and purveyor of military hardware to Kaiser Wilhelm II. Friedrich Alfred ("Fritz") was searching for a new steel alloy that would be harder than cast steel. At the time, the armor of the world's battleships measured two solid feet in thickness at the waterline. Fritz ordered experiments with nickel steel. It worked brilliantly. At a stroke, all existing plate patents and all naval armor had been rendered obsolete. The Kaiser was elated. Fritz advertised it in every chancellery in the world. By the turn of the century he was official armorer in Moscow, Vienna, and Rome. Then Fritz unveiled chrome steel shells that could pierce the nickel steel. Now the chrome steel became the rage. Next, at the world exposition at Chicago, Fritz demonstrated a new high-carbon armor plate that would resist chrome steel shells. Among others, the United States Navy beat the path to the Krupp factory in Essen, Germany. And so it went.

The agony of bringing on line a never ending series of ever more advanced technological products has become highlighted by the efforts the U.S. semiconductor industry. The market leader is Intel Corporation. Intel's 486 microprocessor was introduced in 1989; at the time it was the most powerful chip available. But behind Intel was a pack of competitors, led by Advanced Micro Devices. Three years later, AMD announced the successful "cloning" (=duplication) of the 486. The next generation of microprocessors was the 586 chip, named "Pentium", introduced in late 1993. (The 486 is a 32 bit microprocessor, meaning that it addresses 2^{32} locations of memory. The 586 is a 64 bit processor addressing 2^{64} locations of memory. A bit = binary digit is the smallest logical unit in a computer.) The economic lifespan of each new wave of technology in the semiconductor industry is about three years. That adds up to a lead time of eight or nine years from the start of research to market introduction. Peak output of the Pentium is expected to be reached about the year 2000. And that is not the end. Research is already in full swing for the development of the 256 bit chip generation. At the same time, preparations are being made for entirely new types of microprocessors with even more stupendous capabilities.

To describe this process, we can use the term "attributes" once more. Through research and development, the list of existing attributes of a product is upgraded, and new attributes are added. Each individual product, characterized by a given set of attributes, eventually must become technologically obsolete. It will inexorably end up in the technological museum. Instead, the path of progress moves sideways, so to speak. New superior products with superior lists of attributes replace the old product. The diesel locomotive replaces the steam locomotive. The electric locomotive replaces the diesel. Next there is the lightweight train, the modern commuter train, the Japanese "bullet train," the monorail, the magnetically levitated train. New product cycles overtake the old ones. There is no end to material progress, as long as man is capable of dreaming up new attributes.

Concepts like "technology," "progress," and "economic growth" thus somehow are intricately tied to the diversity or heterogeneity of economic goods and services. Managers perceive a discrepancy between the list of attributes that current technology is able to deliver, and a list of attributes that lies within the realm of possibilities. It is that discrepancy that prompts current efforts at product development and new commercialization. In the final analysis, it is the discrepancy that drives progress. The driver of progress is disequilibrium: the leap between what is and what could be.

What is Productivity?

New technologies are at the heart of increased productivity. In a seminal paper titled "Technical Change and the Aggregate Production Function," published back in 1957, Nobel laureate Robert M. Solow studied the behavior of gross national product (GNP) per man hour in the U.S. over a 40 year period, from 1909 to 1949. During the period, productivity in this sense about doubled, from 62 cents to $1.27 (his calculations were in 1939 dollars). Solow devised a method of splitting this increase into two components: one component due to increased use of capital, and one component due to "technical change." It turned out that 87.5 % of the increase was attributable to technical change, and only 12.5 % to increased use of capital.

Solow's work came to have strong influence on the work of subsequent generations of academic economists. It posed a challenge: what precisely is this technical change that seemed to explain most of the improvement in productivity? In a series of publications, E.F. Denison (1916-1992) tried to provide an answer. He developed a list of factors that could reasonably be expected to determine output, such as employment and hours, education, increased experience and better utilization of women workers, changes in the age-sex composition of the labor force, nonfarm residential structures, other structures and equipment, inventories, U.S. owner assets abroad. Denison proceeded to measure the rate of growth of each such factor, and to relate them to the rate of growth of GNP. This time it was possible to explain a much greater proportion of the growth of GNP. But there was still some "unexplained residual" -- the technical change. In an early study Denison reported (for the U.S.):

Time Period	Rate of Technical Change
1953- 1964	1.68 per cent per year
1964- 1969	1.37 per cent per year
1969- 1973	1.13 per cent per year
1973- 1976	-0.65 per cent per year

(The 1970s were the bleak years of oil crisis, stagnating economies around the world, and accelerating inflation -- the so called "stagflation.")

Next, Denison turned to the spectacular economic performance of the Japanese economy during the post World War II years. Studying the growth rate of GNP in Japan from 1953 to 1971, he was left with an unexplained residual equal to 1.97 per cent per year. Again identifying the residual with the rate of technical change, he concluded that technological change in Japan had been much faster than in the U.S.

But other economists have found this line of reasoning disquieting. Reviewing Denison's work, the outstanding econometrician Richard Stone wrote:

"The existence of a residual [in the estimation of growth] is a great weakness and in my opinion should be got rid of. Every component contributing to the growth or decline of output should be quantified ..."

Certainly, there must be something fundamentally wrong with the priorities of an analysis where the driving forces of change in the economy are seen as residual -- left after all other factors that the economist can think of have been exhausted.

One problem with Denisons's productivity calculations, and the host of numerical exercises in the same vein that economists have carried out, is that it assumes that technological change occurs at an even and constant rate over some time period, such as over each five year period in the table above. That is, technological change comes at hand like some manna falling from the heavens, inducing all statistically measured factors to increase their yields at an even rate.

The Technology Frontier

A quite different approach, drawing upon a different heritage of economic theory, would see technological change springing from unexpected technological and marketing breakthroughs, emerging challenges and opportunities, and a perennial state of disequilibrium. In brief: a theory of uni-directional and jerky shifts.

There is no reason to believe that such shifts in an industry would tend to occur at an even rate over time. Quite on the contrary, there is strong theoretical and empirical support for the assertion that innovations -- large and small -- tend to cluster over time. Initial breakthroughs set the stage for follow-up innovations. Inside a particular field of applications, there may occur a rapid cascading of innovations, such as the recent rapid progress in cellular telephone technology, or biotechnology.

How does new technology raise productivity? It might lie near at hand to believe that the productivity gain sets in as soon as the new technology is introduced. But that is not so. In its opening phase, new technology typically delivers about the same productivity as existing technology. The important productivity gains occur when the old technology is abandoned. What really counts in the national productivity statistics are the closing down decisions, the discontinuation of outdated production methods or of outdated production facilities. Firing workers, idling capacity, closing down factories -- those are the kind of decisions that are apt to break hearts. But the national economy benefits from the eradication of low productivity operations.

In brief, there is strong backing for the assumption that an industry moves from one productivity "regime" to the next at discontinuous points in time, marking each upswing in the industry cycle.

Rather than assuming a fixed growth rate in each five year period, one may then think of the following computational maneuvre: insisting that shifts of technology must be uni-directional jerks rather than smooth time rates, instruct a computer to search through all possible partitionings of the total time span into individual subperiods to find the partitioning that fits the data best. Applications of these ideas leads to the following results for the manufacturing sector in the U.S., 1968 - 1987:

timing of shift (shift from one "regime" to the next)	magnitude of shift (uni-directional jerk)
1968 - 1969	+ 1.49 %
1971 - 1972	+ 3.49 %
1972 - 1973	+ 5.35 %
1983 - 1984	+ 3.83 %
1985 - 1986	+ 0.85 %
1986 - 1987	+ 6.14 %

There were rapid improvements of productivity in the early 1970s. Then come the stagnating years -- the eleven years between 1973 and 1983. During this entire time period the estimated production relation stayed unchanged. Only in the late 1980s did productivity start growing again.

The message is clear. Productivity in the U.S. stagnated in the 1970s. Not until 1983 did productivity start growing again, first hesitatingly, but eventually, in 1986-1987, quite solidly.

Pushing the Technology Envelope

There is another difficulty with the productivity calculations that I have now cited. The reported figures are statistical averages. Some companies enjoy a faster productivity growth than the average, others a slower growth. But the figures that we are after, really, are those at the technology frontier or the leading edge. The frontier is no average. It is the maximum that current technology permits.

The frontier can be determined by a mathematical technique called data envelopment analysis. It was invented in the late 1970s by A. Charnes and W.W. Cooper (see box). The technique can be applied to a wide range of decision-making units, certainly to corporations engaged in manufacturing or in distribution or marketing, but also to not-for-profit bodies such as electric utilities, public schools, and hospitals. The purpose of the calculations is to trace the technology frontier and to rate each individual unit relative to the frontier. Some notable applications reported in the literature involve U.S. army recruiting districts, U.S. Air Force wings, and even Chinese cities!

Data envelopment analysis was brought to the attention of the general public with the successful site selection of the superconducting supercollider to Waxahachie, Texas (south of Dallas). In 1984, the Texas legislature funded a four-university study to identify feasible sites for location of a high energy physics lab in Texas. Six feasible sites were identified and a comparative site analysis was made, incorporating project cost, user time delay, and environmental impact data. The Waxahachie site was found to be preferred under a wide range of conditions.

A company located on the technology frontier is a cost-effective, that is, it generates its outputs with a minimum expense on inputs. Economists call this "efficiency." Companies in economic textbooks are always "efficient." Since the days of Adam Smith and Alfred Marshall, economics has been the science of the behavior of "economic man," some fictitious rational individual who, weighing all possible courses of action, chooses the one that maximizes "utility" or profits or whatever. A widely used definition of economics states that it is the science of allocating given means to obtain a set of desirable ends. But the suboptimal companies are not very good at solving that optimizing problem. And yet, they are very much part of the reality that economists need to come to grips with. There are companies that are not very good at maximizing profit.

A. Charnes and W.W. Cooper
A Remarkable Operations Research Team

In the scientific world, research teams are formed and dissolved as individual research interests evolve over time. But sometimes two researchers will stick together, working and publishing together for a longer time. A. Charnes and W.W. Cooper did research together for most of their scientific lives.

Charnes and Cooper were brought together working on a problem of blending aviation gasolines. This was the first real-life application of a new mathematical technique called linear programming. To solve it, Charnes developed theory of handling so-called degeneracy, and Cooper carried out the hand calculations. It was a classic "first" in the literature of the new science of operations research (OR). It was the beginning of a joint career that would see the publication of half a score of books and several hundred joint research papers.

Charnes (1917-1993) was trained as a mathematician and headed for many years the Institute for Cybernetic Studies at the University of Texas at Austin. Cooper is a student of the great Paul Douglas, a Chicago economist known for his empirical work on production in the 1930s. Cooper has served as the dean of the School of Urban and Public Affairs at Carnegie-Mellon and was a professor in accounting at Harvard before he moved to Texas.

The driving force of Charnes' and Cooper's research has almost always been applications. They were sought after for consultation work by private industry, government and the military. In this way they became exposed to real-life problems. In order to deal with them they developed theory and analysis and numerical methods. To ensure that such developments were relevant and useful, they often included officers from the consulting party into the research team as co-developers and coauthors. For example, the technique of chance constrained programming arose from a consultation project for Esso. A corporate officer of the company had approached Charnes and Cooper seeking help to schedule the production of heating oil in the face of demand that fluctuated strongly in response to the weather. He was also a coauthor of the final research paper.

In this fashion, the body of mathematical and programming techniques of OR has arisen in response to the needs of applied studies. There is a lesson to be learned here. All scientific work must start with a problem. Next you need background and data. Theory is developed later, as the researcher tries to fit the problem in a formal framework. Theory is only useful as far as it helps answer real-life problems.

Data envelopment analysis (DEA) arose in a study of 'Program Follow Through,' a large-scale social experiment in public school education conceived in the late 1960s intended to supplement the pre-school program Head Start. The result of the analysis was a ranking of all participating schools in terms of their achievements (verbal scores, math scores, pupil self-esteem etc.) per unit of resources spent.

The engineers may not see or understand the technological opportunities. The manager may simply be a poor manager. In order to understand the real world, we need a theory of management operating at the cutting edge of technology and we also need a theory of management that falls behind.

The possible existence of "non-economic man," of companies that fall behind the technology frontier, has of course been recognized from time to time. Nobel laureate Herbert Simon speculated that many companies may not minimize costs, but rather choose to adopt a stance of "satisficing behavior," that is, aiming for "satisfactory" rather than maximal profits. Nevertheless, data envelopment analysis has brought about a revolution in economics. For the first time, economists can now measure these matters empirically. Companies located at the frontier are assigned an efficiency value of 1. Units located behind the frontier are given an efficiency rating less than unity.

To illustrate, we may take a quick look at some results for the U.S. computer industry 1980-1991. The calculations are based on standard financial data brought from earnings statements and balance sheets of publicly traded corporations. A computer program gauges the performance of each computer stock relative to the others.

First, a number of indicators of performance (outputs) are chosen, such as total sales, earnings, stock capitalization (the market value of a company's outstanding shares). Other indicators measure inputs into the productive process, such as employment, selling and administrative expenditure, capital investments, expenditure on research and development. While this may seem like a long list, it nevertheless represents only a first cut at the problem. For a more detailed analysis one may want to break down the sales volume by type of computer hardware sold, and its engineering characteristics. This is what financial analysts call the "fundamentals." If one really wants to go to the bottom of the matter, it is not sufficient to read annual reports. One needs to visit factories, talk to executives, and learn the nuts and bolts of the trade.

The envelopment calculations show that some corporations systematically obtained top scores in terms of their productive efficiency. During the decade, the winners were:

Apple Computer	National Computer Systems
Atari Corp.	Quantum Corp.
Compaq ComputerCorp	Seagate Technology
Conner Peripherals	Silicon Graphics
Dell Computer Corp.	Tandon Corp.
Floating Point Systems	

Among these eleven corporations, Apple Computer, Compaq Computer, and Seagate Technology stand out. Apple was at the frontier every single year except 1989. A few of the winners (like Dell, listed on Wall Street since 1987) are recent start-ups and have a much briefer track record.

Other companies were falling behind the frontier. They were not efficient. They did not use best available practice. They represented inoptimal or suboptimal management. The list includes households names in the computer industry such as Amdahl Corp., Data General Corp., Digital Equipment, Hewlett Packard, Sun Microsystems, Unisys, and Wang. IBM had been at the frontier in the early 1980s but fell behind already in 1984.

Ironically, several of the suboptimal companies were featured as showcases of "excellence" of management in the well-known book "*In Search of Excellence*" by Peters and Waterman (1982). At the time, they were singled out as role models for their strategic performance and their ability to adapt themselves to a changing environment. Their management style was widely admired. But in the last resort the competitiveness of a corporation refers to its bottom line, to its cost effectiveness. On that score, these companies fell short.

One of the great stars of the computer industry was Sun Microsystems, formed in 1982. After its introduction on Wall Street in 1985, Sun saw its sales multiply more than twentyfold during the next five years. And yet, Sun was subefficient every single year, falling solidly behind the cost effectiveness frontier. How can that be? The answer, obviously, is that the Sun management consciously sacrificed short term cost efficiency in order to achieve rapid long term growth. SUN was riding the wave of its line of immensely successful computer workstations, built around the Unix software (originally developed by AT&T).

The modern history of computers is the history of a few immensely successful companies, like Apple, Cray, Microsoft, SUN that during the opening phases of their life cycles all grew at unbelievable rates. Such phenomena pose problems for the professional economist who insists on seeing the world through the lenses of the "theory of equilibrium." These are wild divergent disequilibrium processes. These companies revolutionized their industry, pushing the technology frontier to new unseen territory. And yet, as we have seen, during that phase of explosion they may never have been able to move up to the static frontier. The great technology revolutions are disorderly, flaunting the narrow book-keeping concept of cost effectiveness!

The case of SUN is particularly intriguing because it illustrates how a computer company can exploit the chemistry between hardware and software to launch itself onto a trajectory of supergrowth. SUN fervently believed in "open computing," making its software developments (its own version of Unix, and software for local area networks) available to everybody at nominal fees. The result: SUN in effect came to set the standards for the emerging workstation industry, indirectly building a huge demand for SUN hardware and a huge demand for Unix.

So, who is leading the advance into the technological unknown? Is it those companies at the cost effectiveness frontier, or is it those along the second line of assault, shoving and pushing to get ahead? As the reader will realize, the dynamics of the battle is complex. There is no field marshal in command, field tube in hand, astride his white horse, pointing out the lines of attack to his soldiers. It is every CEO for himself. There is only one certainty: those who fall too far behind will falter.

Complexity and Evolution: The Santa Fe Institute

The Santa Fe Institute is an offshoot from the nuclear weapons laboratories and computer laboratories at nearby Los Alamos, offering an opportunity for hard-core physicists and computer jocks to mix with scholars in "soft sciences" like economics and social sciences. The complex problems of the modern world -- nuclear waste, the greenhouse effect, genetic experimentation -- demand expertise over a broad range. Unfortunately, most researchers tend to focus on a single narrow aspect of these problems. In the real world, everything affects everything else. The purpose of the Santa Fe Institute, says founder George Cowan, is to educate "a kind of twenty-first century Renaissance man, starting in science but able to deal with the real messy world."

The institute has become renowned for its work in the study of complex behavior in mathematical, physical, and living systems. In 1986 John Reed, Citicorp's chairman, put up some money and asked the Santa Fe Institute to work on a new, improved model of the world economy.

Discarding the standard notions of equilibrium, economists working at Santa Fe developed a new kind of economic systems with "positive feedback," like real-estate booms and stock market crashes. Once such systems get rolling, they keep snowballing and feed on themselves, up to a point. Positive feedback arises in an economy when there are increasing returns to scale. If the average production costs of some firms fall with increasing output, the managers of these firms will find it profitable to expand their scale of operation. Outputs and market shares of individual firms will embark upon some dynamic path of growth.

At first look, such phenomena may seem to be at variance with standard assumptions of economic textbooks. As a producer increases output, his total costs are usually portrayed as eventually rising quite rapidly. Average costs would then also rise. There are two different things at issue here. On the one hand, there is the theoretical thought experiment of textbook economics, referring to a hypothetical situation of a producer increasing his output while factor prices and technology are kept unchanged. In such an artificial setting it is reasonable enough to assume that unit costs would eventually increase. In terms of economic equilibrium theory, the problem of determining the output and the inputs of a producer has a "stable" equilibrium solution.

In empirical studies it is virtually impossible to control and freeze the technology variables. In a population of companies studied during a given time period, an observed increase in output will statistically go together with improved technology. In a population of companies studied over several years, more advanced technology is employed at later dates. As a result, the statistician will conclude that an expansion of output leads to falling unit costs.

The envelopment analysis described a moment ago presents an opportunity to determine the possible presence of increasing returns to scale. It turns out that the returns to scale tend to vary systematically over the life cycle of a technology. In the beginning of the cycle there will typically be large returns to scale. Young and expanding startup companies enjoy increasing earnings. But as the technology matures it soon needs to be replaced by a later technology generation. If the company lags behind in updating its technology, the company will be stuck in a position with decreasing returns to scale. At that point, the choice is to scale back or to make an all-out effort to regain the technological initiative.

In addition, there are powerful mechanisms reenforcing positive feedback among a pack of competing firms. There is synergy in the development of high technology. Corporations learn not only from their own experience but also from their competitors. Each manufacturer does not have to invent the wheel all over again. He can climb on the others' shoulders, adapting existing technical designs, taking note of what seems to sell for his competitors and what does not sell. There is also synergy in marketing. When one manufacturer launches a marketing campaign for one particular product, he automatically prepares the ground for the next wave of product designs and the next manufacturer coming along.

A rapid turnover of product cycles is self-reinforcing in that it tends to speed up the rate of product development even further. As managers watch their competitors flood the market with new and advanced product designs, they realize that their own brands will become obsolete even faster than originally expected. To maintain their market share, each competitor will need to pour even more resources into research and development, and to accelerate their various projects already under development.

When self-reinforcing mechanisms are at play, there is little reason to expect the resulting growth path to be orderly and linear. Companies will scramble to exploit the opportunities at hand. New start-up companies will enter the fray.

Santa Fe economist B. Arthur has been looking into the possibility that the resulting dynamic economic system can possess a multiplicity of solutions, that is, a multiplicity of alternative future courses. There may exist many candidates for long term self-reinforcement; the cumulation of small random events early on pushes the dynamics into the orbit of one of these and thus selects the time path that the system eventually settles on. Arthur cites the example of the competition between Sony's Betamax and its rival VHS in video technology. Although most experts would agree that Betamax represented a superior technology with better picture quality, VHS got an early lead. There was positive feedback, and eventually VHS was able to capture the entire market. Beta became obsolete.

The price of video recorders and of video movies continues to come down. In the early spring of 1995, Disney in one single week sold more than 20 million video cassettes of its hit movie "The Lion King." In the future, Hollywood may make movies exclusively aimed at the video purchase market (so-called "sell-through" tapes as opposed to tapes for rental). The market is growing by leaps and bounds, in a disorderly and unpredictable fashion.

The result can be a hectic boom, as the boom in the semiconductor market in the early 1980s. That boom eventually crashed as the market became saturated. In the process, many U.S. producers folded and the Japanese strengthened their grip on the market. In a contracting market, the pace of rejuvenation slackens and obsolescence of existing designs does not set in as quickly as before. Managers slow down their product development efforts and delay the commercialization of new entries in the market.

These cyclical mechanisms are not unsimilar to the booms and busts that occur in steel, shipping, and construction. When freight rates are good, shipowners contract a lot of new tonnage. A couple of years later the new tonnage hits the market. The excess capacity depreses freight rates.

But there is a difference. The conventional excess capacity cycle is a cycle in the construction of real capital. The new booms and busts trace a cycle in research and development. We have only limited experience of these new cycles so far, mainly on the positive side. The decade of the 1980s was a period of unparalleled upswing of high technology. I am asking myself with some nervousness what a massive general downswing in the high technology sector would look like. Are we in the process of building into the modern economy an element of cyclic instability that makes the U.S. economy susceptible to prolonged recession, even depression? A greater proportion of the U.S. workforce than ever is occupied in the manufacture of future goods and services. Can this backfire to the present? I do not know.

Outside the undergraduate library of the University of Texas at Austin is a statue that I have admired many times. It shows two runners participating in a race of relays. The leader is an old man. He is strained almost beyond endurance. He is about to collapse. Behind him a glorious young athlete is approaching, catching the baton that is slipping from the hand of the older one.

He is fast. He is determined. He will win. The message of the artist is obvious: a new generation will take over. But I also think of something else. New generations of technology will replace the old ones. What seemed to one generation the ultimate technological perfection, will seem impractical or even foolish to a new breed of engineers and marketing specialists and managers. There will always be progress as long as man is able to speculate, and think, and dream.

Bibliographic Notes

The subject matter covered in this chapter is what economists refer to as "the advance of technology," leading to "shifts" of production functions and of efficiency frontiers. After a few introductory considerations, the text turns to an account of Solow's 1957 estimations of productivity change in the U.S. and subsequent efforts along the same lines by Denison and Kendrick (for references, see below). To these early researchers increased productivity was essentially a residual item -- the increase of output remaining after the contributions of labor and capital had been accounted for. (Kendrick's concept of total factor productivity TFP is defined in this manner.)

Denison's estimations drew on conventional econometric techniques, allowing for the presence of statistical errors. Using the well known method of least squares, the estimation is carried out so as to minimize the sum of all squared errors. The result is a fitted mathematical representation of output in terms of labor and capital. Permitting it to shift smoothly over time (that is, using calendar time as an additional explanatory variable), productivity is obtained as the time rate of such change.

The alternative approach of discrete uni-directional "jumps" of productivity is based on the idea of partitioning the entire time span of analysis into a sequence of productivity "regimes." Productivity remains unchanged during each regime, but is permitted to increase from one regime to the next. The econometric estimation is carried out with the help of so-called dummy variables, one dummy for each regime. The timing of the regimes is effected so as to minimize the sum of all squared errors (see Thore *et al.* below).

Productivity estimation took a new turn with Aigner and Chu ("On Estimating the Industry Production Function," *American Economic Review*, Vol. 58, September 1968, pp. 826-839). Discarding the method of least squares -- that is, essentially calculating productivity as a kind of statistical average -- they argued that the frontier productivity should not be calculated as an average at all. The frontier supposedly represents the optimal performance. Accordingly, the statistical variation around it can only be one-sided: a company may fall behind the frontier, but it cannot surpass it. Aigner and Chu fitted a production function to the U.S. primary metals industry with all observations falling on one side of the fitted relationship.

Economists today distinguish two approaches to such one-sided estimation. One is the "parametric" approach (such as the one used by Aigner and Chu), estimating conventional productivity relationships but requiring the errors to be of one sign only. The other is the "non-parametric" approach. It is the approach of data envelopment analysis (DEA). The envelope to a scatter of production observations determined by DEA is identical to the "efficiency frontier" of conventional production theory. Each point on the envelope is said to be Pareto-optimal, i.e. given the inputs, it is not possible to increase any one single output without having to reduce another output.

References

"Imagine that we could enter today the first exhibition..." J.B. Priestley, *Victoria's Heyday*, William Heinemann, London 1972.

"The passenger train was one of the glorious products ..." I have used N.L. Wayman, *St. Louis Union Station and Its Railroads*, The Evelyn E. Newman Group, St. Louis, Missouri 1987.

"This is the so-called 'learning curve'" L. Argote and D. Epple, "Learning Curves in Manufacturing," *Science*, 23 February 1990, pp. 920-924.

"An early master -- and an evil master -- " W. Manchester, *The Arms of Krupp*, Little, Brown and Co., Boston 1964.

"The agony of bringing on line..." C. Prestowitz Jr., *Trading Places: How We Allowed Japan to Take the Lead*, Basic Books Inc., New York 1988, and J.George, "High-Technology Competition between U.S. and Japanese Companies," in *Japan Business Study Program*, 1989, ed. by H. Matsuo, Bureau of Business Research, The University of Texas at Austin.

"As a convenient point of departure..." R.M. Solow, "Technical Change and the Aggregate Production Function," *Review of Economics and Statistics*, August 1957, pp. 312-320.

"In a series of publications, E.F. Denison tried to..." E.F. Denison, *The Sources of Economic Growth in the United States and the Alternatives Before Us*, The Committee for Economic Development, New York 1962, *Why Growth RatesDiffer*, The Brookings Institution, Washington D.C. 1967, Accounting for United States Economic Growth 1929-1969, The Brookings Institution, Washington D.C.1974, *Trends in American Economic Growth, 1929- 1982*, The Brookings Institution, Washington D.C. See also E.F. Denison and W.K. Chung, *How Japan's Economy Grew so Fast*, The Brookings Institution, Washington D.C. 1976. For other key works on productivity, consult J.W. Kendrick, editor, *International Comparisons of Productivity and Causes of the Slowdown*, American Enterprise Institute, Ballinger Publishing Co., Cambridge, Mass. 1984 and Angus Maddison, "Growth and Slowdown in Advanced Capitalist Economies: Techniques of Quantitative Assessment," *Journal of Economic Literature*, June 1987, pp. 649-698.

"The existence of a residual..." R. Stone, "Whittling Away at the Residual: Some Thoughts on Denison's Growth Accounting," *Journal of Economic Literature*, Dec. 1980, pp. 1539-1543.

"How does new technology raise productivity?" For an instructive statistical investigation, see K.-O. Faxén, C.E. Odhner, and R. Spånt, *Lönebildningen i 90-talets Samhällsekonomi* (Wage Formation in the Economy of the 1990s), Raben & Sjögren, Stockholm 1988, pp. 173-176.

"Application of these ideas leads to..." S. Thore, "A Constrained Least Squares Method for Estimating the Effects of an Unknown Monotonically Intervening Factor," *Journal of Forecasting*, Vol. 8, 1989, pp. 369-379 and S. Thore, G. Xia and T. Song, "Irreversible Technological Progress and Innovations: Endogenous Estimation of Discrete Shift Variables," presented at the Georgia Productivity Workshop, The University of Georgia, Athens, October 1994, submitted to *Journal of Productivity Analysis*.

"The technique is called data envelopment analysis..." A. Charnes, W.W. Cooper and E. Rhodes,"Measuring Efficiency of Decision Making Units," *European Journal of Operational Research*, Vol. 3, 1979, pp. 429- 444. Also, by the same authors, "Data Envelopment Analysis as an Approach forEvaluating Program and Managerial Efficiency - with an Illustrative Application to the Program Follow Through Experiment in U.S. Public School Education," *Management Science*, Vol. 27, 1981, pp. 668 - 697.

"... with the successful site selection of the super-collider to Waxahachie, Texas" R.G. Thompson, F.D. Singleton, Jr., R.M. Thrall, and B.A. Smith, "Comparative Site Evaluations for Locating a High-Energy Physics Lab in Texas," *Interfaces*, Vol. 16, 1986, pp. 35-49.

"To illustrate, we may take a quick look at ..." The DEA calculations mentioned in the text are reported in S. Thore, "Cost Effectiveness and Competitiveness in the Computer Industry: A New Metric," *Technology Knowledge Activities*, Vol.1, No.2, Fall 1993, pp. 1-10. See also S. Thore, G. Kozmetsky, and F. Phillips, "DEA of Financial Statements Data: The U.S. Computer Industry," *Journal of Productivity Analysis*, in press 1994.

"The purpose of the Santa Fe Institute, says founder George Cowan. . . " Cited after M.M. Waldrop, *Complexity: The Emerging Science at the Edge of Order and Chaos*, Simon & Schuster, New York 1992, p. 68.

"Santa Fe economist B. Arthur..." W.B. Arthur, "Competing Technologies, Increasing Returns, and Lock-In by Historical Events," *The Economic Journal*, March 1989 and "Positive Feedbacks in the Economy," *Scientific American*, February 1990, pp. 92-99.

CHAPTER 5

Economic Logistics

The Greek word *logistikos* simply means "calculatory," "rational." When the word entered the Latin language, however, the meaning changed. The main use for calculations to the Romans occurred in the military field, for the counting of the number of legionnaires, swords, chariots, and horses. And so, to the Romans, the word came to mean the procurement, distribution, maintenance and replacement of military materiel and personnel. That is the meaning of the term military logistics today.

War presents enormous logistics problems. Superior logistics planning can very well determine the outcome of war. The Imperial general staff in the Kaiser's Germany, under the command of Schlieffen, had calculated the movement of every train load, every field gun, and every company of soldiers to the front for the planned war with France. Once the order of mobilization was given (day M), everything would follow like a clockwork. On day M+1, the soldiers would embark, on day M+3 they would arrive at the front, on day M+4 the forward penetration of the enemy lines would be initiated in Belgium. On day M+39, Paris would fall.

General Pershing, who in 1917 was placed in command of the U.S. forces, faced an even more difficult logistics problem: starting from scratch, with no war plans at all for the European theater, with little materiel of any kind, no air force, a small fleet, and only the territorial army, he was to transport two million men over the Atlantic ocean, unload them in French ports that were hopelessly underdimensioned for such a task, build depots, railway lines, and to move his troops through France to the battle front in the Ardennes.

Operations research was developed during World War II as first the British and then the American headquarters called on mathematicians and statisticians and other scientists to develop methods for dealing with problems of tactical and strategical planning. The efforts of these first OR teams contributed to the successful outcome of the Air Battle of Britain and the war in North Africa. Other mathematicians studied the deployment of aircraft for bombing raids over Germany, and submarine warfare.

It was recognized at an early stage that many of the mathematical techniques of OR had commercial applications. One of the first was inventory control. A retailer operates a warehouse. It is costly to store goods in the warehouse (there are financing costs, labor costs, insurance charges, and perhaps even theft); on the other hand it is no smart policy to do without inventory either, because many customers who are not served right away will cancel their order. So, there exists some optimal inventory. The early operations researchers came up with mathematical formulas for the determination of an optimal inventory policy over time. Eventually, with the advent of the electronic computer, it was possible to construct computerized systems that solve numerically for reorder points (if the inventory on hand falls below the reorder point, the system will automatically trigger a reorder of inventory).

Other early civilian applications of OR included transportation networks, scheduling of machine operations, and sequencing. The term "management science" became established for the commercial application of such OR techniques.

The next milestone was the founding of ORSA, the Operations Research Society of America, and TIMS, The Institute of Management Sciences, in 1954. ORSA covered the military field, and TIMS the commercial applications. (The two societies were recently merged.) Both societies are concerned, among other things, with logistic systems. TIMS regularly sponsors research on multi-echelon inventory systems, hierarchical production and distribution networks, and the optimal location of plants, inventories and retail outlets. In the bylaws of the TIMS College on Logistics, logistics is defined as

"The study of systems and structures of the provision of people, materiel, and services to organizations."

As examples of economic logistic systems, take some time at your breakfast table to contemplate the paths that the various products have travelled before they reached your table. The tea bag: tea leaves grown in Assam, harvested, transported to warehouses in Dacca, packed, loaded on freighters, shipped to New York, processed into tea bags, distributed via one of the nation's large supermarket chains. The orange marmalade on your toast. The strip of bacon. The muffin. Or, if you prefer a healthier diet: the bran flakes, the grapefruit, the rye bread. Behind each product there is a network of production, storage, transportation, and distribution, often spanning half the globe. It is one of the marvels of the capitalist economy that everything that you would care for (and is willing to pay for) actually is routed to your table in a timely and orderly fashion.

The Dimensions and Mathematical Techniques of Economic Logistics

A logistic system has three dimensions: the spatial or regional dimension, the vertical flow of a commodities from the initial use of resources, via the processing of intermediate goods to the completion of final goods demanded by the consumer, and the time dimension. To illustrate the nature of these dimensions, I propose to discuss three mathematical and computing techniques developed by operations research and management science. The first is network analysis, highlighting the regional dimension. The second is activity analysis, representing the vertical dimension. The third is multi-echelon (or hierarchical) inventory systems.

First, networks. A network consists of links and nodes. In the Trans-Siberian pipeline, the links are the pipes, extending from the Siberian tundra to residential and commercial consumers in Germany, Holland, Belgium, France. The nodes are the control stations and forking points along the way. In the American Airlines flight network, the links are the various individual flight connections and the nodes are the various airports served. In the AT&T telephone network, the links are the cable trunk lines, domestic and overseas and satellite channels, and the nodes are the switching stations.

Imagine the task confronting an airline officer who is to draw up the time table for the scheduling of all passenger aircraft. Given the expected number of passengers arriving at each node wanting to be transported to various destinations, and given the fleet of aircraft, it is a well-defined mathematical problem to construct the optimal utilization of each aircraft, providing maximal service to customers (with minimal waiting time) and minimal costs. Or suppose that you are in charge of production at a large integrated oil company, pumping oil in North Africa, shipping it to the U.S. in tankers, and distributing it in all 48 contiguous states. The oil company may own some of the tankers outright, and lease others. It may own port facilities and storage tanks in several cities along the entire coastline of the U.S., and refineries both in North Africa and in the U.S. It owns a fleet of trucks for the short-haul transportation of gasoline. Given all production and transportation costs, and the demand at various delivery points, which is the optimal routing of the flow of crude oil and gasoline? Again, this is a well-defined mathematical problem. It can be solved by network theory.

In order to solve the network problem numerically, a computer has to be given information about all the routing possibilities and the attendant cost along each link. The computer will then search for the optimal routing of the flow along the network. Such search proceeds step by step, by a so-called algorithm. The basic idea is simple. First start by an initial "feasible solution," any feasible routing scheme meeting the demand for transportation. Next, the computer will be instructed to search for step-wise improvements of the initial routes. In large networks, there may be hundreds or thousands of such intermediate steps before the computer arrives at the final and optimal route.

OR specialists have constructed algorithms that successfully can deal with immensely large networks -- with thousands of nodes and thousands or tens of thousands of links.

Second, activity analysis. The concept of activity analysis was introduced in economics by Tjalling Koopmans, a Dutch economist who had settled at Yale and eventually got a Nobel prize for his work. He shared his price with the

Russian Leonid Kantorovich who a decade earlier, in Leningrad during the war years, investigated the optimal combinations of individual industrial processes or activities in local manufacturing plants.

In a plant, or a farm, or a hospital, the operation of production at the efficiency frontier typically involves the concurrent use of several alternative technologies. For instance, there exists in the radiology department of a modern hospital a great number of different imaging techniques that the doctors need to consider. Each procedure requires its own expensive equipment and servicing and back--up personnel. Some may be carried out in the hospital, others may be subcontracted to specialized medical imaging firms. The management of the hospital then faces the mathematical problem of determining the right mix of those technologies. That problem was solved by Kantorovich and by Koopmans. It is called "activity analysis." The solution: management will choose to operate some optimal bunch of activities. Other feasible activities are not operated. CBS operates both radio stations and TV stations. The radio stations transmit on both the AM band and the FM band. It owns some stations outright, and has exclusive agreements with others. It sells news programs overseas.

In a logistics system, activities are concatenated together in a production chain that extends all the way from the original use of labor and natural resources (so-called primary goods) via the processing of intermediate goods to the ultimate delivery of final consumer goods. In the case of the teabag that you use in your morning tea, a bale of packed tea leaves loaded in the port of Dacca is an intermediate good. So is a carton of tea bags handled in a wholesale warehouse. The final good is the package that you buy in your retail store.

The World Bank through its Development Research Center is promoting the use of mathematical sector models and industry models, using various programming formulations of activity analysis. Notable such models were developed for Mexico and the Ivory Coast. Other studies have dealt with the steel industry, the fertilizer industry, the forest industry and so on. In order to facilitate the mathematical programming involved in these studies, the World Bank took the initiative of constructing a new software language called GAMS (= general algebraic modeling system).

Third, multi-echelon (or hierarchical) inventory systems. To a manufacturer distributing a product over a large geographical area it is rarely smart to store the inventory in just one single warehouse. Instead, the product may first be stocked a regional warehouse, and later transferred to a local warehouse serving a local area. Department stores like Sears and Montgomery Ward maintain hierarchical systems of warehouses and storage depots. If you order a piece of furniture at your local outlet, it will be routed to you through a sequence of warehouses. I have already characterized the nature of the mathematical problem of determining the optimal inventory in one single warehouse. The mathematics becomes more complicated when there are entire echelons of warehouses. There is then not only the task of managing each warehouse separately, but also the wider problem of the location and the optimal capacity of each of the various warehousing units, and the flow of goods between them.

In a computerized hierarchical inventory system, computers at the various warehousing sites may be wired together enabling them to communicate with each other, monitoring the flow of goods through the entire warehousing network.

As explained, the science of economic logistics employs mathematical and computer techniques that were originally developed for the analysis of a single firm or a single industry. Economic logistics extends the use of these tools to study fundamental questions concerning the allocation of scarce resources in an entire economy, the choice of technology, and the satisfaction of consumer demand. Just as the chemist in his laboratory can create polymers like nylon with thousands of atoms combined in a single molecule, we realize that it must be possible -- at least in principle -- to build a vast mathematical model involving thousands or millions of details, spelling out all the manufacturers, all intermediate goods, all inventories and production facilities, and the routing of all transportation in an entire industry.

I am not necessarily suggesting that one should undertake this Herculean task. Rather, what I am after is conjuring up a particular vision in the mind of the reader, a theory if you will, of the structure of the economic system. This vision involves a network of concatenated manufacturing, transportation and distribution activities extending along the entire production chain from the use of natural resources to the delivery of final consumer goods. In principle it would be possible to spell out the details of this system and to represent them mathematically.

Mainstream economists will not feel comfortable in such a far-flung theoretical world. Economists like simplicity. Economics as taught in college is "micro" economics and "macro" economics. Microeconomics is the behavior of consumers and producers. Macroeconomics is the economics of aggregates -- national income, gross national product, and total demand. Economic logistics is a tool that makes it possible to chart the abyss between micro and macro. One reason why economists have scored so few successes in their attempts at explaining what goes on in the real world is that they prefer to reduce complex processes to the simplistic language of aggregate theory. Economic logistics faces the complexity of the real world head on.

The formal apparatus might require pages and pages of mathematical text just to write it down even in condensed mathematical notation. But here the OR analyst gets some unexpected help. The electronic computer can do what the human mind cannot: to develop, store and remember mathematical structures of forbidding complexity and magnitude. The computer can be used not only to process data. It can also be used to develop and store theory -- theory that is so vast that it becomes impossible for a single human mind to keep track of all the details. It is possible to develop these things in the memory of the computer because of the existence of high-level computer languages that by simple commands can create new links or new nodes in a network.

Innovation and Technological Change in an Industry Network

How can technological change be represented in a network of industrial activities? An innovation may be a new technology to manufacture or distribute or market a known good -- a new link in the network. It may be a new consumer good, manufactured by a new technology -- a new link and a new node. Or it may be a new producers' good, manufactured by a new technology and used as an input into another new technology -- a new link and a new node and a new link. So, more generally, innovations involve the appearance of new links and new nodes, sometimes minor additions to the existing network, sometimes major changes.

A macro-innovation, like the telephone or the movies, introduces an entire new arm of technologies in the network, branching out into hundreds or thousands of new detailed products and manufacturing processes.

When new technologies emerge, old ones often succumb. When telegraphy across the U.S. became a reality, there was no longer any need for the pony express. When jet propulsion of airplanes became commercially feasible, the propeller engine lost a large portion of its market. Nobody goes to German health spas any longer to "take the waters." The reason is not that these places are less charming than they used to be. There may still be a band playing in the pavilion in the town park of Baden Baden on a balmy summer night. The reason is the advance of medicine and our understanding of how medical afflictions might best be treated.

The economic mechanism is this: In order to motivate the innovation, the new product or the new technology has to acquire some piece of the market. It can only do so by offering a price low enough to drive out some existing product or technology. Tourist trips to the moon will become a commercial possibility when the price is low enough to induce some potential customers to choose the moon rather than a cruise in the Caribbean. As customers switch, the market for cruises will weaken, and some marginal operators of cruise ships will have to close down.

In a rapidly advancing economy, there will always be pockets of decline and pockets of unemployment. If markets are free, workers are willing to move, and the price of machinery and plants and land is permitted to fall, the freed-up resources will eventually be reemployed again. To Joseph Schumpeter (1883 - 1950), the discontinuation of obsolete technology actually was part of the creative process itself. He called it "the gale of creative destruction." Out of the tumultuous confrontation between several competing innovators, less efficient technologies are weeded out and a higher kind of evolutionary order is established.

The old has to make way for the new. Bankruptcy, the shutting of factory doors and the firing of workers from inefficient production units serves an important economic function. That is of course one of the reasons why technological change is slow in under state socialism.

In order for creative destruction to work, it is necessary that workers and engineers and computer programmers and nurses are willing and able to retrain or reeducate themselves so that they can move from the subefficient workplace to an efficient one. In many cases, it may be a momentous and traumatic experience to labor. An entire local community may be hit if the town plant closes down, an automobile assembly plant or a steel mill. Younger workers may be able to leave, to retrain and to find a new job. But what about the older ones? If labor cannot move, it becomes all the more important that the price of capital is permitted to fall. If the price of the deserted plant falls sufficiently, somebody out of town may find some use for it. A closed-down shipyard may be rebuilt into a factory manufacturing prefabricated housing. A closed-down railway station may be converted into a shopping mall (Union Station in St. Louis). A Tulsa, Okla. oil company cleaned up the lining of an old oil pipeline, inserted a fiberoptic communications network into it, and got into the telecommunications business.

The capitalist system provides a chance for everybody. But not all will succeed. Bankruptcy and discontinued production means lost equity capital. Money is lost. But private capitalism is for the common good. This, of course, Marxists have never understood. Society benefits when productive resources are moved out of lines of business that yield a loss.

Progress and growth occurs as the logistics network grows in all directions, like a salt crystal in a saturated solution or a colony of amoebas being grown in a test tube. New links and new nodes are being added, in all directions, all the time. In jumps and starts, entire new branches form and send off off-shoots. Old branches wither away. It is a living organism, multiplying and becoming more and more complex.

Budding and Cross-Fertilization

The emerging growth pattern, then, is one of organic growth, as in the growth of a living system. To emphasize this biological analogy, economic growth may be discussed under two headings: budding and cross-fertilization.

Budding occurs as the creation of new links and new nodes in the logistics network breeds additional links and additional nodes incident upon those already in place. Such growth is "hierarchical." The adoption of new technologies sets the stage for subsequent product development and innovation. As one vintage of technology is succeeded by a superior one, it will in its turn eventually be superseded by even more advanced technologies. New technologies are raised on the shoulders of earlier achievements.

Most of the early product development in the automobile industry can be characterized as budding, such as the honeycomb radiator, mass production, the self-starter, lacquer finish, and the assembly line.

Cross-fertilization in a technology network occurs when two or several nodes located in different parts of the network (both presumably having been established through processes of local budding) are joined together through the creation of new links (and new nodes). Whereas budding is the archetype of product development and gradualism, cross-fertilization represents the leap into the unknown, the dramatic technological discontinuity, the true innovation, like the fax machine (cross-fertilization of digital data technology, telecommunication, and laser printers), commercial satellites launched from airplanes (cross-fertilization of rocket technology, aviation, and satellite technology), the artificial heart (cross-fertilization of electro-mechanical engineering, blood technology, and surgery).

Although cross-fertilization in the technology network typically leads to dramatic change, such change always occurs in an orderly and structured fashion, preserving the sub-hierarchies that are merged together via the new links and the new nodes. The introduction of commercial overnight mail services (Federal Express) required the setup up of a combined system of customer mail pickup and delivery, the operation of aircraft, and establishing a national airport hub. All these separate technologies existed before, but were now coordinated into a single packaged product, and sold via aggressive marketing. A new logistic hierarchy was put in place.

Cross-fertilization may lead to entire strings of new links being set up, bridging nodes that seemingly had only a distant relevance to each other. Deep-sea oil drilling has brought us the drill-ship, a curious combination of maritime shipbuilding and drilling and seismic exploration technology. The exploration of outer space has spawned entire industries engaged in the manufacture of mechanical robots, the utilization of solar power, the development of new ceramics and alloys, rocket technology, and electronics. But all these new technologies and new products can easily be identified by their sub-technology ancestry. Innovation from scratch, as it were, is extremely rare. (Cold fusion, if proven feasible, perhaps will provide an example.)The Striking of Oil in the Neutral Zone and the Launching of Cable News Network

Technological uncertainty arises because most technologies cannot be fully controlled by man. There are intervening random elements, like the weather and plant diseases in agriculture. Even if a farmer plans very carefully what kind of crops he is to grow on various pieces of land, deciding on methods of irrigation and cultivation, on the use of agricultural machinery etc., the output will still be stochastic.

There has been rapid technological development in almost all phases of the production and distribution of seafood during the last decades, spurred by breakthroughs in marine biology, breeding technology, packaging and air freight. But these advances are fraught with uncertainty. An example is provided by the crayfish industry. The state of Louisiana accounts for 85 % of the world's harvest of crayfish (called crawfish in Louisiana). The state's Atchafalaya Basin is a vast wetland, where the crayfish finds an ideal natural habitat. In addition, industrial breeding in ponds has become important and during the last decade the pond acreage in Louisiana has tripled. The crayfish pest is caused by a parasitic fungus. It can rapidly kill the entire population of crayfish in a breeding pond. Thus, the production process is quite risky. Distribution is also precarious, since the crayfish is a perishable commodity. Through packaging and air transportation it is possible to reach distant export markets.

Modern product development can be planned and managed. But it must at all times respond to ongoing R&D work, test marketing, and evaluation. It is not possible to determine in advance the set of new product designs that are going to be marketed 18 months hence. You cannot lay down a master plan for the required results of technological development and make it stick. There is technological uncertainty.

How do decision-makers act in the face of uncertainty? A common hypothesis in economics is that less uncertainty is preferred to more uncertainty, all other factors being equal. As applied in the conventional theory of stock portfolios, this maxim leads to the conclusion that portfolio managers will select a balanced portfolio providing some optimal trade-off between return and risk. Similarly, one may introduce the concept of a "technology portfolio," arguing that firms will seek to establish a balanced portfolio of uncertain production prospects to secure some optimal trade-off between return and risk. In any line of manufacturing or trade there will exist some optimal mix of well-established production techniques and the development of new ones. No hospital will just do heart by-pass surgery. There will exist an optimal mix of activities in each operating room, blending well-established surgical procedures with some experimental ones.

The technology portfolio will cover some reasonable band width in the technology spectrum. A manufacturer of small propeller-engine aircraft will make pleasure airplanes, executive airplanes, airplanes for use in agriculture (spraying pesticides), forestry (watching for forest fires), the oil industry (checking for leaks in pipelines). It will not make helicopters, nor jet aircraft. A manufacturer of sports shoes will make tennis shoes, basketball shoes, jogging shoes, even martial arts shoes. It will not make boots for trout fishing. Each manufacturer will assemble and develop technology that covers some related product niches where the company can unfold its particular expertise.

Using the language of portfolios, we can view the entire logistics process from the supply of resources, via production of intermediate and final goods, to the marketing and selling to consumers as a vertical chain of technology portfolios. There is a technology portfolio at each link along this chain. Consider shipbuilding and shipping. Shipyards hold and develop portfolios of construction technology, like computer-aided ship design, drydocks, computer-aided cutting of steel plate, mechanized assembly of ship sections, automated inventories etc. Shipowners hold portfolios of freighters, tankers, supertankers, ships designed for the transportation of liquid natural gas or other chemicals, cruiseships. There is uncertainty at each link. The construction of tankers may have to be entirely revised in the light of experience gained from oil spills and environmental disasters. The operation of cruiseships in the Caribbean is vulnerable to political turmoil in the area. And so on.

A good example of vertically linked technology portfolios is provided by the semiconductor and computer industry. Chip manufacturers are racing to develop ever more advanced chips. Computer manufacturers are locked in a fierce struggle to develop new PC's, workstations, peripherals. Computer software and consultant firms are developing new attractive packages for spreadsheet analysis, visual designs, word processing, and financial analysis. Notice in particular how the technological uncertainty at any one link in this vertical chain depends on and is conditioned by advances being made (or expected to be made) at other links: The competition in peripherals prompts new efforts in chip technology, the development of new printer technology prompts the development of new software etc. etc. These interdependencies run both upstream and downstream the logistics flow.

The analytical concepts of technological uncertainty and a chain of technology portfolios can be used to explain the nature and concept of innovation and technological progress. In brief: Innovation and progress occurs as a result of reduced technological uncertainty. When an innovation occurs, there will typically take place a rerouting of the flow of goods and services through the economy. New links will be set up in the logistics network, and existing links will disappear. One innovation triggers a series of others. There is a cascading of innovations along the logistics chain. Eventually, an entire new industry may be spawned.

Using instructive terms, we may say that innovations occur as a result of a "clarifying of technological uncertainty." As information about new product development efforts and test marketing becomes available, the fuzziness of the technological prospects is reduced. Some alternatives, which earlier had to be reckoned with, and contingencies set up to deal with such alternatives, can now be discarded entirely. Other alternative outcomes, which earlier seemed only barely possible, become more probable. There are examples everywhere. A drug company embarks upon the production of a new drug once the results of medical experiments have been confirmed and the new drug has been approved. The approval confirms that the risk of possible harmful side effects is sufficiently low (a drug against AIDS not being excessively toxic, the carcinogenic effects of Nutrasweet and saccharine being below required threshold levels, etc.) The jet age in commercial aviation arrived with the British Comet aircraft. But in 1954 two Comets disintegrated over the Mediterranean. Engineers realized that metal fatigue -- a phenomenon new to aviation -- was to be blamed, and learned eventually how to deal with it. Four years later Boeing's 707, Americas first jetliner, was successfully put into service by PanAm.

A highly instructive case is provided by the striking of oil in the Neutral Zone (located between Saudi Arabia and Kuwait) by Paul Getty in 1953. After having drilled for 3 years and sunk close to 30 million dollars in the project, Getty had still not found a single drop of oil. In March 1953 in the zone known as "Wafra No. 4," oil was struck at a depth of 3800 feet. There was "clarifying of technological uncertainty." From that moment on, Getty set in motion a massive program of investment and expansion. In Wilmington, Delaware, bulldozers were levelling a five-thousand-acre site bordering the Delaware river, in preparation for the construction of the largest refinery in the world. Dredgers were cutting a deep ship channel, three miles long, to accommodate the tankers. In the shipyards of St. Nazaire, France, the keel of a huge new tanker was being laid -- the first of a fleet that would move crude from the Persian Gulf to Wilmington. In the Neutral Zone, Arab laborers were laying pipelines across the desert to link the oilfields with storage tanks being built on the shore of the Persian Gulf. A submarine pipeline was laid to the deep-water anchorage offshore, where the tankers would take on their cargoes.

(Together with all other Getty oil assets, the Wafra oilfield eventually ended up in the possession of Texaco Inc. During the Gulf war its wells were damaged by fire. The refinery was destroyed. But, in early 1992, Texaco let it be known that oil production was resuming.)

For another example, involving the cross-fertilization of several different technologies, consider the case of the launching of the Cable News Network (CNN). Back in 1976, Ted Turner, a self-made TV tycoon, had started sending the signals of his Atlanta "Superstation" up to a satellite hovering 22,000 miles high. Returning to earth, they could be received by the fledgling cable-TV market over all of North America and beyond. Turner's next idea: an all-news program beamed to the world. From a technological point of view, it would merge TV, satellites, cable, and international news operations.

As events were to show, the riskiest link in the proposed logistics system was the satellite. Turner had leased one of the twenty-four transponders on the new SATCOM III, owned by RCA American Communications. The satellite was launched by NASA in December 1979 aboard a three-stage Thor-Delta rocket. The firing of the rocket went without a hitch. Four days later, the ground-station engineers lost contact with the satellite. It was never recovered. The "down-links" were also risky. There were nearly five thousand different cable systems in the country at the time, but most of them took a wait-and-see attitude. At the debut less than two hundred cable systems had signed up, translating into an audience of a pitiful 1.7 million viewers.

The news operations were to be directed from the headquarters in Atlanta. To get the news, CNN scrambled to set up local news bureaus in Washington D.C., New York, Chicago, San Francisco, Los Angeles, Dallas, London, and Rome. Hundreds of staffers had to be hired and trained, and paid a salary, while the new company had not yet made a single dollar in revenues. The up-front costs were big enough to make or break Ted Turner. He had put everything he owned on one card. The rest is history. Ten years later, the CNN coverage was being received by more than half of all homes in the United States, and in eighty-four countries. The CNN success stories include live broadcasts from Tiananmen Square in Beijing (with a million Chinese demonstrating for democracy) and from Baghdad (reporting during the air attacks). The world would never be the same again. The global village had arrived.

The risks very much include possible counter moves by competitors and new entrants into the industry. CBS, ABC, and NBC each tried to buy CNN. Ted Turner made an attempt to buy CBS. Westinghouse and ABC scheduled a competing all-news network with a repeating headline service; to forestall them, Turner started his own CNN Headline News. The players in the market are all the time jockeying for improved strategic positions, looking for new technological opportunities and ways to reduce the awesome technological risk.

Bibliographic Notes

This chapter reviews material that I have developed formally and mathematically in Ref. [18]. The concept of economic logistics is similar to the well-known idea of military logistics, but is expanded to include such features as the optimal location of plants, inventories and retail outlets, and the management of hierarchical multi-echelon production, inventory, and distribution systems. The study of individual features of this supply process is familiar from operations research. They are joined together here into larger structures encompassing the production and distribution system in an entire industry.

The subject matter of Schumpeterian "creative destruction" is mentioned, but, by and large, I do not find this concept very helpful in explaining the dynamics of the high tech economy. It does not tell us much about the nature of the "creative" part of the process. Instead, to discuss innovation in logistics networks, I have proposed the concepts of "budding" (new nodes with their connecting links emerging from adjacent nodes) and "cross-fertilization" (new links straddling nodes located in different parts of the network). See the reference below.

The idea of "clarifying" of expectations is conventional and can be found in F.H. Knight (*Risk, Uncertainty and Profit*, Houghton Mifflin, Boston, Mass, 1933) and in Svennilson and other authors of the Swedish school during the 1930s.

References

"Once the order of mobilization was given (day M)..." B. Tuchman, *August 1914*, Constable & Co., Ltd, London 1962.

The TIMS College on Logistics..." C.H.Hanks, "Of Strong Wills, Dedication to a Discipline, and Lofty Goals: TIMS's College on Logistics," *OR/MS Today February* 1990, pp. 30-35.

"OR specialists have constructed algorithms that successfully can deal with immensely large networks..." See for instance F. Glover, R. Glover, D. Klingman and N. Phillips, "A New Polynomially Bounded Shortest Path Algorithm," *Operations Research*, Vol. 33, 1985, pp. 65-73.

"That technique is called linear programming. It was rediscovered in the West..." G.B. Dantzig, "OR History. Impact of Linear Programming on Computer Development," *OR/MS Today*, August 1988.

"Notable mathematical models were developed for Mexico and the Ivory Coast..." L.M. Goreux and A.S. Manne, *Multi-Level Planning: Case Studies in Mexico*, North-Holland, Amsterdam 1973 and L.M. Goreux, *Interdependence in Planning. Multilevel Programming Studies of the Ivory Coast*, Johns Hopkins University Press, Baltimore 1977.

"...the World Bank took the initiative of constructing a new software..." A. Brooke, D. Kendrick and A. Meeraus, *GAMS. A User's Guide*. The Scientific Press, Redwood City, Calif. 1988.

"Out of the tumultous confrontation between several competing innovators, less efficient technologies are weeded out..." In addition to the classical references to Schumpeter's works (above), see R.U. Ayres, "Technology: The Wealth of Nations," *Technological Forecasting and Social Change*, Vol. 33, 1988, pp. 189-201.

"To emphasize this biological analogy, economic growth may be..." S. Thore, "Innovation in an Industry Network: Budding, Cross-Fertilization, and Creative Destruction," *Honoring Gerald L. Thompson on the Occasion of his 70th Birthday*, Kluwer, in press 1994. The distinction between incremental and discontinuous innovations is commonplace, see the overview in W. W. Rostow, *Theorists of Economic Growth from David Hume to the Present*, Oxford Univ. Press, New York 1990, Chapter 20.

"Using instructive terms, we may say that innovations occur as a result of a 'clarifying of technological uncertainty'..." S. Thore, "Dynamisk Preferensteori Under Osakerhet" (Dynamic Preference Theory under Uncertainty"), *Ekonomisk Tidskrift*, No.2, 1962, pp. 37-73.

"After having drilled for 3 years and sunk close to 30 million dollars in the project, Getty had still not..." R. Miller, *The House of Getty*, Henry Holt and Co., New York 1985.

"...consider the case of the launching of the Cable News Network..." H. Wittenmore, *CNN: The Inside Story*, Little, Brown, and Company, Boston 1990.

CHAPTER 6

Of Princes, Diplomacy, and Battle Cries

The features of capitalism that I have described up to this point cannot have disturbed many readers. High tech and a continuous flow of new product designs hitting the market place is attractive to most people. In essence, what I have dealt with is "free enterprise" as it occurs toward the end of the twentieth century. Now I come to a much more controversial subject matter and I shall run the risk of offending many. Free enterprise is not the same as capitalism. Free enterprise is a system of management. But capitalism, at its heart, is a system of ownership and power: private ownership and private power. The subject matter that now confronts us is the exercise of that power.

To some people, the misuse of corporate power is legendary. There exists an underbrush of political pamphlets and campaign rhetoric that wants us to believe that corporate profit is based on deceit, that workers are "exploited" and that the consuming public is offered poor quality goods that it neither wants nor needs. Corporations are seen to neglect worker safety, flaunting government regulation. Multinationals are being accused of exploiting poor third world countries. To the well-known list of accusations a few new ones have been added: the destruction of the environment, and toxic and nuclear hazards. All these terms and concepts are based on concepts of misuse of power. And because capitalism is the system of individual economic power, capitalism as an order of organization of the productive means in a society is seen as being tainted too.

So, just how is individual power exercised under capitalism? How has it been exercised, and how is the exercise of power changing in the era of high tech capitalism? Is it evil or is it for the common good?

In his book *The Prince* published in 1513, Niccolo Machiavelli set out to study the behavior of the princes of his day and to formulate the rules for success, including the use of arranged marriages, treachery, deceit, and even murder. At that time, Italy was fractured into a mosaic of dukedoms, city states and ecclesiastic states, each ruled by a "prince" -- some princes of the blood, others sheer adventurers. Warfare was the order of the day, one prince seeking aggrandizement at the expense of eviction of another. Ferrara, Padua, Pisa and other cities were involved in a never-ending power struggle, sometimes captors, sometimes captives, just like Paramount Communications, the target of an epic takeover battle in 1993-1994 fought between Viacom (with brand names such as Nickelodeon, MTV, Blockbuster), and QVC Network.

I would like here to embark on a similar undertaking, investigating the exercise of corporate power in our time, laying down the rules for corporate success, be they morally palatable or not. My aim is not to judge. It is to understand.

Sometimes peaceful negotiations could do the trick. Francesco Sforza signed a marriage contract to prevent the Habsburgs from taking his dukedome in Milan. The appointed bride was only eleven and a half years old. The duke did not even bother to show up for the marriage ceremony. One of his emissaries represented him, exchanging the matrimonial vows with the young bride.

But if diplomacy and strategic alliances did not work, one had to resort to war. Machiavelli exhorted the prince of his day to

"... have no other object, nor any other thought, nor take anything else as his art but the art of war and its orders and discipline; for that is the only art which is of concern to one who commands."

If the present-day corporate raider is the prince of the 14th and the 15th century, then certainly the yuppie is the knave, fighting under the banner and bounty of the *grande seigneur*. Once a knave needed physical prowess, wielding the lance or the battle-axe. Their modern counterparts, crowding the corporate stables in lower Manhattan, armed with PC's and attache cases, need similar stamina, working long hours, sacrificing their own comfort, and ready to receive deadly body blows. It is the fight itself, the macho display of daring force and the facing of danger, that keeps their spirits aglow.

The Strategies of Takeovers and Acquisitions

I now propose to examine the diplomacy and the wars of the captains of corporations. Diplomacy takes the form of strategic alliances. Wars are fought for takeovers and acquisitions.

The first strategy which is as old as capitalism itself is to buy up competing companies so as to establish a monopoly and to be able to control prices. The second strategy is vertical integration -- buying up companies along the entire production chain so that you control the flow of a product all the way from natural resources to the final consumer product. Rockefeller built Standard Oil into a world integrated oil company, controlling the flow of oil from the oil fields to the pump.

These two were the standard strategies of the captains of industry before 1950 -- in coal, steel, railways, movies, chemicals and so on. They belong to an earlier era of capitalism when the development of new technologies was comparatively slow, markets were static, and it was possible to build enduring monopolies. But even if it is difficult to reach and maintain a market share above, say 60 %, in today's world, corporate warfare with the purpose of expanding your market share can still pay off. Aggressively buying out competitors in a rather narrow product line where the rate of technological advance is comparatively slow, it is still possible to become a major player in one's market. This is the model that some of the European "transnationals" have chosen, such as Electrolux (refrigerators), Nestle (chocolate), and Asea BrownBovery (electric transformers and generators). For the recipe to be successful, the expanding company needs to establish itself as a market leader in R&D as well, building defensive product development capabilities against the possible assault from new upstart competitors. But the days of a static and enduring market dominance are certainly gone.

The dangers of vertical integration in the face of rapidly changing technology must be pointed out. Vertical integration means that plants and distributors which are positioned downstream the production chain are committed to buying their inputs from units under the same corporate aegis. If those upstream units do not manage to stay abreast in the technological race, the downstream units will base their operations on component technology that is falling behind.

General Motors and many other U.S. manufacturing corporations are engaged today in a process called "outsourcing." It is the very opposite of vertical integration, phasing out the manufacturing of components and parts inhouse and buying them instead from Mexico or other low-cost nations. If you look under the hood of your Buick or Oldsmobile, you will spot parts emanating from all over the world. Ostensibly, the purpose is to reduce costs. But the reason is also technological: the assembly plant keeps its options open, always standing ready to switch the purchases to the manufacturer offering the technically most advantageous product.

The third strategy is broadening the product line -- horizontal integration rather than vertical integration. There are often considerable economies to be found in gathering the manufacture and marketing of several related products under one management. Economics textbooks call it "economies of scope." To a firm manufacturing and selling a household cleaner it is probably a good idea to buy a detergent company and to enter the detergent market as well. Once you possess one successful brand name, the consumer will recognize your brand among similar products displayed in the same aisles in the supermarket. When Disney built its theme parks Disneyland and Disney World it was capitalizing on the name recognition of Mickey Mouse and Donald Duck and all the other well-known figures from the animated movies. When Disney added Epcot center it capitalized on the name recognition that the earlier theme parks had gained. And when Disney expanded its hotel operations in Orlando, it was the same kind of horizontal expansion by means of consumer name association. The new product complements the former ones. Thus a corporation may grow organically through acquisitions and new ventures, its product line expanding to related areas. Perhaps the next natural step for Disney is to buy an international hotel chain (it already built a "Euro Disney" at Marne La Vallee outside Paris). Or to buy a cruise line (to be based in Long Beach, say, where the company already docks its Queen Mary and the Spruce Goose). One thing is certain: management must be on the move, if possible one pace ahead of the consuming public. It must make acquisitions. There is much to be said for a "friendly" takeover. But if the target company does not go along, a hostile move must be considered.

There are other even more compelling reasons for dynamic management of the product line. As current technology and markets are continuously evolving, a firm needs to disengage itself from technologies and products that are subject to rapid technological obsolescence and to pursue alternatives that hold a greater promise for the future. Consider the case of Eastman Kodak. For many years this company dominated the home movie industry. But that industry -- as we now know-- was headed for extinction. Recording pictures by means of photographic film will eventually disappear entirely. To survive, Kodak must move into the field of electronic recording and storage of images. To accomplish what is needed for a company of Kodak's size, it is not just enough to start an electronics division from scratch. It is not easy to enter a new technologically demanding field where you have no experience. The way to do it is to buy an existing operating electronics company specialized in the field where you want to go -- in this case, electronic images and video disks. You need to buy up a going concern. You need to wage war -- to select your target, and go after it. Once a controlling interest has been acquired, the acquisition must be digested. You need to move in your own corporate officers, turn the research and development of the new unit toward your own goals, and to shed those parts that you do not need. (Kodak acquired Interactive Systems, a Santa Monica, Calif. computer firm in 1988. The former head of that company was put in charge of Kodak's electronic imaging business. And, in 1993, Kodak made an even more dramatic move: raiding electronics giant Motorola, it it named George Fisher, the electronics maverick who had built Motorola into a semiconductor and communications giant, its new chairman and executive officer.)

The management of AT&T had realized for a long time that in order to stay in the lead in the telecommunications business, they needed to develop a strong data division. In markets such as automatic teller machines and retail terminals, transactions are generated by the millions each day. They are fed directly into local and long-distance communications networks. At an early point, AT&T had developed the UNIX operating system, a computer software enabling different kinds of computers to "talk" to each other. AT&T built its own computers and sold them under its own brand name. But in spite of these efforts, the AT&T data division was a mess, running up huge losses. And so, in 1991, AT&T management decided that there was just one solution: to buy in the marketplace the data networking expertise that it had failed to build in-house. They set their eyes on NCR (National Cash Register Co.), a company that for more than one hundred years had built shiny brass cash registers and which in the 1980s had achieved impressive results in data processing in banking and retail. In a hostile bid, AT&T bid $90 a share for the NCR stock. The NCR management was not amused. NCR's defenses included a "poison pill" that blocked AT&T from acquiring shares without NCR board approval. The NCR executive officer, Charles Exley Jr., adopted what he called "the Greta Garbo defense: we want to be left alone." But when AT&T raised its bid to $110 a share, Exley relented.

Sometimes diplomacy can achieve just as much as war. Two years later the aggressive chairman of AT&T, Robert Allen, made his next move, buying McCaw Cellular Communications.

The fourth strategy is risk diversification. Corporate management faces risk all the time. Some of these risks are of an engineering nature (such as the Comet aircraft disasters over the Mediterranean in the early 1950s, or the Challenger explosion). Accidents occurring in the production process can entail environmental hazards (Three Mile Island, Bopal). The risks confronting an integrated oil company include the possibility that a skipper of an oil tanker drinks a glass of whiskey too many, that the tanker runs aground, and that the ensuing oil spill rings up billions of dollars in cleanup costs and punitive damages (Exxon, 1989). The risks confronting a distributor of a major consumer household product includes the risk of an alarming find of a chemical contaminant (Tylenol, the Alar scare, Perrier).

With the general public becoming aware of the relationship between ultraviolet radiation and skin cancer, the development, production, and marketing of tanning oil with sun screen factors has become fashionable. As the medical knowledge of these matters increases, current technology can easily become obsolete overnight. There exists a strong incentive for a manufacturer to branch out to neighboring technologies, say cosmetics, skin cleansers etc. involving well-known technologies with low risk.

There are also political risks, risks that plants or installations in a foreign land are seized by a hostile government. There is the risk of debilitating labor unrest and strikes.

What can a corporation do to manage risk? The answer is diversification -- diversification of technologies and diversification of the product line. And the fastest road to diversification is corporate acquisitions. In the wake of the oil crises of the 1970's, when oil prices peaked and the large integrated Oil companies were reaping extraordinary profits, many observers thought it incomprehensible and downright unpatriotic when some oil companies used this infusion of cash to invest in entirely different lines of business rather than drilling for more oil (case in point: Mobil Oil bought Montgomery Ward). But these companies were already- - as events had shown -- heavily exposed to risk. They needed to lower the overall risk content of their technology portfolios. (Most of these acquisitions were later disgorged again, with the oil companies returning to their core line of business.)

Major Corporate Takeovers in the U.S. since 1985			
Year	Target	Acquired by	Purchase sum $ billions
1985	RCA	General Electric	6.4
	Shell Oil Co	Royal Dutch Shell	5.7
	General Foods	Philip Morris Inc.	5.6
	Hughes Aircraft	General Motors	5.0
1986	Beatrice	Kohlberg Kravis Roberts	6.2
	Safeway Stores	Kohlberg Kravis Roberts	5.3
1987	Standard Oil	British Petroleum	7.6
	Southland Corp	Acquisition group	5.1
1988	Kraft Inc.	Philip Morris	13
	Federated Dept Stores Corp.	Campeau Corp.	6.5
	Farmers' Group	Acquisition group	5.2
	Sterling Drug	Eastman Kodak	5.1
1989	RJR Nabisco	Kohlberg Kravis Roberts	25
	Columbia Pictures	Sony Corp	5.0
1990	Warner	Time Inc.	14.1
1991	NCR	AT&T	7.9
	Contel	GTE Corp.	6.2
1993	McGaw Cellular	AT&T	12.6
	Medco Containment	Merck	6
1994	American Cyanamid	American Home Products	9.2
	Blockbuster Entertainment	Viacom	7.9
	Paramount	Viacom	9.6

What happened afterwards? Under the leadership of KKR, RJR Nabisco carried out a series of debt restructurings and regained its financial health. Besides building cigarette plants in Eastern Europe, RJR has bought food companies in the U.S., Brazil, and Mexico. In 1991, KKR sold a 60 percent stake in RJR to the public in an initial public offering. Four years later, KKR shed its last holdings in RJR.

Of Princes, Diplomacy and Battle Cries

Safeway Stores went through a major downsizing program, shedding low-profit supermarkets. Four years after the leveraged buy-out, the KKR group sold the company back to the public again. *Eastman Kodak's* purchase of Sterling Drug (including the North-American manufacturer of Bayer aspirin) put unwelcome financial strain on the company as it was laboring to establish itself in the electronic imaging business. In 1991, Kodak was ordered to pay $900 million to Polaroid Corp in a record patent infringement award, having brought out instant cameras back in 1976. Three years later, Sterling Drug was disgorged.

Several of the acquisitions went sour. *Federated Department Stores* (Bloomingdale, Lazarus and many others) and *Allied Stores Corp*, both acquired by Canadian developer Robert Campeau, made the largest retail bankruptcy filing in U.S. history. Four years later, Federated was on the war-path again, merging with Macy's to form the nation's largest department store company. For Sony, the purchase of *Columbia Pictures* turned out to be a huge mistake. There was a clash of cultures, and the company was mismanaged from the start. In 1994, Sony decided to take a $3 billion write-off on the movie studio -- one of the largest losses in Japanese corporate history.

Economists differ between "insurers" and "risk lovers." Only few people are risk lovers, enjoying the danger for its own sake. Corporations certainly are "insurers." They take on research and development projects in order to reduce the total risk content of their portfolio. RJ Reynolds poured millions of dollars into the development of a new "smokeless" cigarette. Taken by itself, it was a highly risky project. But it was not embarked upon in order to gratify anybody's urge for a thrill. Neither was the project driven by any concern about the health of smokers -- management would not have cared. Within the context of the total activities of the company, it was seen by management as a pure act of insurance, spreading the total risk. Most automobile manufacturers experiment with an electric car. Energy companies look into alternative energy sources, such as geothermal energy or shale oil. Aircraft manufacturers look into the possibility of ultrasonic passenger traffic. In this book I am exploring the creative urge that drives such efforts, and thousands like them -- the surge toward the technology frontier. But corporations are not run by engineers alone, driven by a quest for technological excellence. The chief financial officer is driven by prudence. He, too, will look for new advanced technology. If he can find it in-house -- fine. If it has to be acquired from the outside, on the corporate battle ground, so be it.

The point is that in today's climate of rapidly changing technologies, there is no "conservative" stance left. Just holding on to the old line of business, manufacturing and selling the product that sold well ten years ago, will in all likelihood be the riskiest alternative of all. Management needs to make acquisitions and takeovers to reduce the total risk. To the prince, war is the prudent alternative. The real danger is to retract behind the barricades of a withering product line.

Consider the formation of Time Warner, the media empire, the last big deal of the 1980s. Time Inc. was a publishing giant, a purveyor of the written word, disseminated through its 23-title stable of magazines including People, Sports Illustrated, and Time itself. But what future is there for the written word in the electronic age? Perhaps magazines, newspapers, and books are soon going to go the way of the horsedrawn buggy and the doodoo bird. The prospects of the publishing industry are uncertain. There is risk -- product risk. To deal with it, the management of Time decided to diversify into the new electronic media.

Warner Communications seemed to offer precisely what Time needed: video and movies. Warner had bet on videos, MTV, and Nickelodeon when cable was still in its infancy. Warner had produced Madonna, "J.F.K.," and "Batman." It owned the world's largest record business.

The talks between the two companies started in a friendly manner. A merger was being discussed -- a consolidation of the two companies but no issue of new debt. But then a black prince appeared in the clearing of the forest, ready for a hostile takeover. Warner was not the only movie house around casting about to find a publishing partner. Paramount Communications was another. To salvage their original plans, Warner and Time had to restructure the deal. Time bought Warner outright, piling up new debt to the tune of $ 11.4 billion.

Among princelings, personality tells more than ancestry. The late Steve Ross, the flamboyant CEO of Warner, ended up as the top dog of new company. In three decades he had built a Manhattan funeral parlor into a global entertainment conglomerate. This was the moment of his apotheosis.

How do you select your takeover target? It is a kind of portfolio problem, far more complex than the task of putting together a stock portfolio. Management must engage in a dynamic process of building a portfolio of manufacturing and distributing units. The target company must possess the product know-how and track record of successful product development in the field that the aggressor is after. In addition, the distribution system and the marketing experience of the target may also be desirable. If not, the aggressor corporation will shed these portions of the target and distribute and market the product along his own existing channels. So, what the acquiring corporation is looking for is technological and market potential that complements current operations. And, of course, one must not put all eggs in one basket. A variety of moves must be considered providing contingencies for a wide range of possible future technological scenarios. All in all, it is safer to go for a broader product line than a more specialized one, and to develop several possible technologies for a product rather than just one.

There are dangers in trying to cover too broad a product line, too. Management easily loses detailed product knowledge and insight in the changing technology affecting the manufacturing and distribution of too many products. Conglomerates have not in general turned out successful. The theory behind Litton Industries and other conglomerates of the 1960s was that once you have assembled a dynamic and efficient corporate staff of management -- an "organization" -- then you can run anything, be it movie studies or cement factories. Maybe so, as long as you have a Tex Thornton (Litton Industries) or an Arnold Hammer (Occidental Petroleum) at the helm. There are natural management prodigies who can successfully assume the leadership of any kind of corporation. But such people are not common. Most conglomerates of the 1960s (ITT, Gulf+Western) eventually ran into management difficulties. Many of them were dissolved. Some were butchered up during the takeover years of the 1980s.

Capital Gains: How to Make Them

In order to continue my argument, I need to explain the concept of "capital gains." Corporations merge, or acquire new units, or go after other corporations in the pursuit of capital gains. They shed units in the attempt to avoid capital losses. A capital gain on real estate occurs when the property increases in value. It is realized if the property is actually sold and the sales price exceeds the book value (usually the historical cost price plus the value of improvements and additions). But if the property is not sold, there is an unrealized capital gain that may be put equal to the price that the property would probably fetch minus the book value. The important thing to notice is that capital gains are appreciation on capital -- a revaluation that occurs on the capital account rather than on the income account. Capital gains are not income. (The capital gains tax is not a tax on income. It is a tax on the enhanced value of capital.)

The way to become rich in this world is to make capital gains. Few people become rich by saving out of income. As a young vice president rises inside a corporation and sees his income climb to a few hundred thousand dollars a year, he will typically adjust his life style correspondingly, moving to a more expensive neighborhood, joining a prestigeous golf club, buying a ski lodge at Aspen. But stock options are something else again. Or, better still, start your own company, see it flourish, and sell it. Capital gains are the name of the game if you want to build a private fortune. The same goes for corporations. Cash flow is not enough to build a big and mighty corporation. Few corporations become rich by investment. To make the megabucks, the corporation needs to make capital gains.

How does a corporation make capital gains? By developing new products or new production processes, or tapping new markets. In one word: change. Such gains are unrealized because they have not been converted into cash. They would be realized only if management were to decide to sell the plant or the division or the product brand where the capital gain has accrued. But such realization would in most instances be premature. The full potential of the change may not yet have been seized. Furthermore, the buyer -- if any -- might not see the same potential as current management. Capital gains run up precisely because they involve the application of exclusive knowledge that is not available in the market at large. In the first instance, the capital gain is just a perceived opportunity. As management successfully pursues the opportunity, its manufacturing or distribution or marketing operations appreciate. But the unrealized gain is all the time in the eye of the beholder.

The stock market provides a daily gauge of how the management of the corporation is doing. The market capitalization of a corporation equals the total stock outstanding multiplied by the stock price. According to financial theory, the market capitalization can be viewed as the discounted sum of all expected future profit streams. Banks and trust funds and pension funds employ thousands of financial analysts spending their days weighing the prospects of individual corporations on a golden scale. They are familiar with all available statistics. They know plants and products and managers and workers. They interview researchers and financial gurus. They listen in to expert assessments and to rumors. And their employers, paying their salaries, act on their advice, buying and selling stocks by the billions of dollars, every day. As a result, the U.S. stock market has developed into a wonderfully sensitive piece of machinery, gauging at every hour of the trading day the future prospects of every publicly traded corporation in the land.

The stock price provides a gauge of all outside information available to the stock market. (I shall get to inside information immediately.) As information about successful change hits the stock market, the stock price appreciates. The stockholders in their turn make capital gains. If they choose not to sell and to hold on to their stock, those capital gains are unrealized. If they sell, they are realized. So, the causal train of events is this: A corporation introduces changes in technology, manufacturing processes, administration, or marketing that are favorably received by the stock market. To management, the corporation has made unrealized capital gains. At the same time, the stock price appreciates. The stockholders make capital gains. Theoretically, the appreciation in value of the corporation equals the appreciation in value of its stock.

Now for the inside information. Management, presumably, is all the time ahead of the outside information. It spots opportunities and draws up plans -- inside information. As management acts to pursue these opportunities, initiating new research projects, developing new products, reorganizing manufacturing processes and distribution, it will see subjective capital gains and possibilities of future gains that have not yet been discounted by the stock market. There is a discrepancy between the inside and the outside information. Hopefully, the capitalization of the firm in the eyes of management exceeds the capitalization in the stock market.

Inside information that is acted upon and implemented by management will by its nature eventually become available as outside information. As plans are executed, their ramifications become obvious. The inside information dissipates and becomes generally available. In the process, that discrepancy between the capitalization of the firm in the eyes of management and the capitalization in the stock market is gradually wiped out. Capital gains perceived by successful management will be translated into a rising stock price. Or, management blunders will be translated into a falling stock price.

As the reader will understand, I am not sure that current law prohibiting insider trading is economically well motivated. (I am not talking about the legalities of the matter, but solely in terms of facilitating the capitalistic process.) Trading on inside information acts to equalize inside and outside capital gains. The prohibition of insider trading will in many instances hinder the smooth dissemination of inside information to the domain of outside information.

Corporate executive officers watch their stock quotations intently. Many are transfixed by their stock price. A CEO who sees his stock price go up, has provided what the shareholders expect from him. His power consolidates. In a very real sense, he presides over a prospering corporation. Economic values have been created -- capital gains -- that did not exist before. Standard measures of corporate performance, such as the price earnings ratio or the price cash flow ratio, are improved. The leverage of the firm, in terms of total debt divided by total market capitalization, has fallen. The credit worthiness of the company in the eyes of bankers and financial analysts has been enhanced.

But a faltering stock price is a sign of imminent danger. The market capitalization is being eroded. The bankers may call, asking for more collateral. Suddenly, the company becomes vulnerable to takeover threats. To any potential aggressor, the total market value of the company has fallen. It is a cheaper buy. As a matter of fact, many U.S. corporations spend great effort and considerable funds to bolster the price of their own stock. In the 1980s stock buybacks financed with long-term debt became a common practice. But such exercises easily become futile. The total market capitalization (= price times number of shares outstanding) should not be effected. A capital loss that has already accrued cannot be made to disappear through financial tricks.

It is important to understand that the capital loss serves a useful function in the economic life of the nation in that it prepares the way for a redirection of management. Perhaps the doubts of the stock market can be relieved by an in-house management shakeup. If not, the stock price will continue to fall and the stage will be set for a takeover. New leaders, new ideas, and new vision will be brought in.

Conventional economic theory sees economic growth as being driven by savings, and investment. But that is only part of the story. Growth is being redirected and amplified all the time by capital gains and capital losses. And mergers and acquisitions are the catalysts of such change. Of course, real investment do take place in the growth sectors of the economy: the construction of new plant and installations, and the purchase of machinery and other equipment. In the high tech age, some of this equipment is extremely sophisticated and very expensive. And in order for a nation to be able to invest, somebody must have saved (unless we borrow from abroad). But if optimistic expectations are prevailing (spurring the act of investment in the first place), the total capital formation will exceed the amount of investment poured into the industry in question. There will occur a simultaneous appreciation both of existing capital and of newly installed capital. An electronics company may install $100 million in new machinery and equipment. The market capitalization of the company may increase by $200 million. There are capital gain multipliers of a successful investment.

Conversely, in a stagnating industry there may occur a rapid capital destruction that eclipses the cessation of net investment. Net investment equals gross investment minus depreciation. In theory that depreciation should include all possible capital losses due to obsolescence or softening markets. In practice capital items are written off over some reasonable time span, like ten years or twenty years. But the reality of the matter is that in the fast-changing world of high technology, capital can be obliterated over night. There are capital loss multipliers of investments that go sour.

A land developer and a corporate raider perform quite similar functions. The land developer buys land with commercial potential. He sets aside some land for a shopping mall, and some land for a residential development. He installs water and sewage, and constructs main road arteries. The value of these investments are enhanced by accompanying capital gains. The developer sees no use for a marshy tract, and decides to get rid of it. Another developer attracted by the low ask price, snaps it up and converts it into a waste dump.

The key elements of a successful corporate acquisition can be explained in the following manner. The management of a company spots an opportunity to make an acquisition, friendly or hostile. Management draws up a plan for the digestion of the acquisition: Some portions of the company to be acquired will be beefed up by net investment. In the process, the acquiring company will make (unrealized) capital gains. Other portions of the spoil will be cashiered.

The management of the acquiring company (the corporate raider, if you will) thus performs the following useful functions to society as a whole: It spots opportunities for capital gains in market capitalization, that is, opportunities for growth. If successful, it seizes those opportunities, transferring control and corporate power as required. Furthermore, it converts unrealized capital losses in trailing production activities into realized ones, thus accelerating the closing down of unprofitable lines of industry.

The Anatomy of a Hostile Takeover: RJR Nabisco

I now turn to the subject matter of leveraged buyouts, a remarkable innovation in the field of corporate transactions. A flurry of such buyouts took place in the 1980s. To illustrate the mechanisms of a buyout, it is instructive to take a look at the largest acquisition ever: the takeover of RJR Nabisco by Kohlberg Kravis Roberts & Co in February of 1989.

Nabisco (= National Biscuit Co.) was created during the turn-of-the-century trust era. At the time, it was referred to as the "biscuit trust." It started baking its Oreo cookies in 1913, and its Ritz biscuits in the midst of the depression. Enter the prince: Ross Johnson, of Standard Brands (Planters peanuts, Fleischmann's margarine, Chase & Sanborn coffee). Johnson was on his way up in the business world; Nabisco was stagnating. The two companies merged to form Nabisco Brands and Johnson became CEO of the new company. Next came the eerie "cookie wars" of 1982 and 1983. Frito-Lay had developed a new soft cookie, Grandma's. Proctor and Gamble unveiled a soft line of Duncan Hines cookies. In the opening skirmishes, Nabisco was blooded badly. But Johnson marshalled his forces and eventually was able to bring his own soft entry, Almost Home, successfully to the market. Nabisco recovered to emerge as one of America's great food companies.

The idea to merge Nabisco and RJ Reynolds Tobacco Company was born in the executive suite of RJ Reynolds. Richard Joshua Reynolds had arrived in Winston Salem, North Carolina in 1874, attracted by the best tobacco-growing land in the country. By the turn of the century, he had built a nationwide distribution chain for pipe tobacco. In 1913 he took his gamble on a new product: the cigarette. He called his brand Camel. In the early 1950s Reynolds Tobacco Company introduced Winston, a filtered cigarette, and Salem, a menthol cigarette. They all became favorites of the smoking public. These were the good times. Then nemesis struck. The surgeon general issued his landmark report linking cigarette smoke with cancer.

The managers of RJ Reynolds realized that it needed to diversify in order to reduce the total risk content of their portfolio of products. The first attempt at diversification nearly wrecked the entire company. Reynolds bought an oil company and the shipping company Sea-Land, building the latter into the world's largest private shipping line. But the new oil and shipping people knew little of the tobacco business. The cigarette brands started faltering under the onslaught of competitors like Marlboro (manufactured by Philip Morris). There was a management shake-up, and in 1984 RJ Reynolds sold its interests in oil and shipping.

The new management of RJ Reynolds set up a task force of staffers and representatives of the company's investment bank to find a more suitable acquisition. They decided to look for a food manufacturer. Reynolds already owned Del Monte; linking up with a major player in food it would be able to demand more and better shelf space in supermarkets and deeper discounts from wholesalers. The list prepared by the task force was topped by Nabisco Brands. Reynolds executives placed a phone call to Ross Johnson. They met in Manhattan the following week. In the spring of 1985 the two companies merged, to form RJR Nabisco, America's nineteenth largest industrial company. Technically, it took the form of Reynolds buying Nabisco for $4.9 billion (a "friendly" takeover). Ross Johnson became the number two man at the helm. A year later he jockeyed himself into the top spot.

Enter prince #2 of this story: Henry Kravis of Kohlberg Kravis Roberts & Co. Or rather -- the King, the undisputed king of the "leveraged buyout" (LBO). Kohlberg Kravis Roberts, an investment firm on Fifth Avenue in New York, had been in the LBO business from the start. Stock prices on Wall Street had been depressed throughout the 1970s. Many stocks were drastically undervalued. These were golden times to buy. As the conglomerates of the 1960s shed businesses, the investment bankers saw the opportunity to buy the cast-off divisions. It was cheaper to drill for oil on Wall Street than in Texas or Oklahoma. Many corporations had liquid assets that exceeded their market capitalization. In this financial climate, investment bankers like Kohlberg Kravis Roberts found a fertile ground for a new idea: the management of undervalued corporations would buy out the stock from the current stockholders and thus acquire the corporations that they already headed. The purchaser would be a shell company set up by the investment banker, who would take a controlling equity position and distribute the remaining stock among the existing management. Since the shell would be saddled with a huge debt, only the most profitable parts of the acquired business could generate the necessary cash flow to meet the interest payments. The banker would sell off less profitable units, and close down unprofitable ones. The butchering accomplished, the capital gains accumulated by the slimmed down corporation would accrue to the new private owners. The investment banker would sell out, cash in his gain, and move on to a new customer. There would be immense fees paid to merchant banking companies, brokers of finance, and legal advisers.

One day in October 1988, Ross Johnson met with his board and proposed a friendly buyout at $75 a share, or $17.6 million in all (the stock had never traded higher than $71), financed by Shearson Lehman Hutton. The plan was to sell off Nabisco and Del Monte. The core of the money-making machine -- the tobacco business -- would remain. Two factors seem to have motivated Johnson: his wish to redeem his stewardship of the corporation in terms of providing the stock owners with a neat capital gain, and his own greed. The following Monday Henry Kravis announced a $90-a-share hostile tender offer for RJR Nabisco. In the subsequent bidding war, the price climbed to $109, or $25 billion in all. Kohlberg Kravis Roberts & Co. emerged victorious. Ross Johnson was without a job, and a new chairman and chief executive of RJR Nabisco was brought in. RJR's European food unit was sold. About 2000 employees were laid off and costs were cut aggressively. But by and large, the company remained intact.

Are Corporate Raiders Good or Bad for Industry?

In the accelerating tempo of technological change and new market opportunities there is a legitimate need for managerial shake-ups, a rerouting of managerial talent and risk-willing capital. Through mergers and acquisitions, the control of corporations passes to new ownership and new management groupings who may perceive technological possibilities and market possibilities that the earlier owners did not see.

Several of the much-heralded LBOs of the 1980s later went sour. With the benefit of hindsight, we know today that management and the financial community greatly underestimated the risks of piling up large debts. The "LBO-mania" of the 1980s is gone. But the LBO no doubt is here to stay.

For all the various reasons that I have enumerated and discussed, corporations need continuously to engage in battle. The control of productive units needs to be transferred so that the executive power at all times rests in the hands of the most visionary corporate leaders. In an important sense, the U.S. economy can be viewed as the sum total of the visions of its CEOs. Restrict their ownership and their control to some historic division of markets and some historic division of power, and one will have curtailed the potential of the entire economy. Or, let their visions clash in battle, and there will occur a shakeout and control will be placed in the hands of those who serve the nation's economy more efficiently.

The mathematical theory of game theory can be used to analyze these corporate battles. Game theory was developed in 1944 by John von Neumann and Oskar Morgenstern. The game that confronts us here is "infinite" or "continuous," i.e. each player of the game has available to him an infinite but countable number of moves or so-called pure strategies. The game is being played by a great number of participants; however, the decisive exchanges will typically take place directly between a few "coalitions" of players. In any given case, one or several coalitions may be formed by the managers, by the stockholders, and by the bankers. How is the outcome of the battle determined? The stock market is the umpire. The game is a "general-sum" game. (It is not zero sum.) The payoff function of the game is the sum total of the market capitalization of all corporations in the nation.

Assume for a moment that no corporate battles were to be permitted. Restrain the moves of all players to those that can be effected within the context of the existing power structure. The game then becomes a "constrained game."

The set of permissible pure strategies is restricted. The maximal payoff of the restricted game cannot exceed that of the unrestricted game. Typically, it will fall short of the payoff of the unrestricted game.

In any case, control will at all times fall in the hands of the coalition who is able to maximize the market capitalization of a firm. In the constrained game, only coalitions within the existing power structure are permitted. But an even greater payoff can be attained if coalitions are permitted that extend beyond the existing power hierarchy and beyond the vision of the current managers.

Under nineteenth century capitalism, power in society was typically vested in a few leading families, the Vanderbilts, the Morgans, the Mellons. Under communism, power is held by the commissars and the political elite. Under high tech capitalism, power is acquired by people who have greater visions than others. There is no need to belong to an old family, or to possess inherited wealth. There is not even a need to possess self-made wealth (although it does help). There is no need to wield political power. The only thing a candidate ultimately needs is the ability to perceive possibilities of manufacturing, product development, or marketing that others have not seen, and to be able to convince his creditors of the soundness of his vision. As long as others put their trust in him, they will put him in the swivelchair of the CEO, they will lend him money, and the power is his to wield. The moment he lets those people down who believed in him and the trust evaporates, his power is gone. It will be transferred to somebody else.

The explosive growth of high technology in the 1980s went together with turmoil in the stock markets. In my estimation, the latter was a precondition for the former. The fighting spirit of the corporate princes benefits -- nay, invigorates -- the capitalistic system. It is on the battle grounds of these corporate tournaments that obsolete technology is discarded and way is made for the new.

It is up to society at large to lay down the rules of battle. Here much remains to be done. If unprotected, many innocent by-standers will get hurt. As hostile acquisitions bring in new management teams, intent on paring costs, the new managers will pay little respect to established routines and practices. They may be outsiders to the industry and strangers to the local community where the company conducts is business. In their zeal, they may violate fundamental ethical rules of business. Wanting to boost production, one corporate raider may callously expose the workers to increased accidents in the workplace. Another raider may start fiddling around with the company's pension fund. A third may pooh-pooh industry safety standards (Eastern Airlines). A fourth may decide to liquidate ten centuries worth of ancient redwood forests (Maxxam Inc.).

The solution to these problems and others like them is not to single out raiders or to make takeovers difficult. It is to enforce the laws of the land in the areas of safety at the workplace, environmental protection, and so on. Modern capitalism only works to our benefit inside a framework of societal regulation, the rules of the game being applied evenhandedly and squarely to established corporations and to new ones alike.

Bibliographic Notes

An early study of the economics of acquisitions, mergers and industrial concentration is Edith T. Penrose, *The Theory of the Growth of the Firm*, Blackwell, Oxford 1959. Penrose argued that a business firm should be seen as essentially constituting a collection of productive resources: its physical resources, its administrative setup, and the entrepreneurial competence of its management. At any given point in time, there exists an inherited structure of such resources. Among these, there will exist a pool of unused productive services and special knowledge providing an opportunity for some internal expansion and growth. But if large opportunities for profitable expansion are seen by management, while the rate of internal expansion is limited by existing productive resources, the solution is external expansion through acquisition or merger.

Inferring from the experience of the Western world up to the time of her writing, Penrose assumed that there are clear economies of scale available to larger firms. But even so, she argued that one would always find in an industry smaller firms operating side by side with the large ones. She pointed out that increasing market diversification and technological specialization presents possibilities which a small firm may more easily exploit than a large firm. She called such opportunities for small firms the "interstices" in the economy. Seizing such opportunities, some small firms would embark on rapid paths of growth. To overcome the limits to expansion set by their pools of internal resources, they would grow by acquisition or merger.

As the discussion in the main text turns to the stock market, the concept of capital gains is explained. Unrealized capital gains typically first accrue in the imputed markets for corporate assets. Only later, as the intents of management become clear to the investing public (i.e. the inside information becoming disseminated to the outside), may such capital gains be translated into unrealized capital gains of common stock. In Schumpeterian "growth blocks" of corporations or industries, unrealized capital gains are mutually reinforcing each other, setting the stage for increased capital spending and increased R&D. (See B. Carlsson and R.G.H. Henriksson, *Development Blocks and Industrial Transformation: The Dahmenian Approach to Economic Development*, Industrilitteratur, Stockholm 1991.)

Toward the end of the chapter I offer a few comments on the use of game theory to study coalitions taking the form of mergers and takeovers. The key feature of the "game" of the stock market is that it is played by managers and stockholders that have only incomplete information about technologies and market potentials. Based on these perceptions one would be able to construct, at least theoretically, a "payoff" function -- a flow of future revenues and costs associated with each production activity. Each player will enter into coalitions and make moves designed to establish a minimax solution: the maximum payoff that he can accomplish given that the other players simultaneously try to minimize it.

References

"..just like Paramount Communications, the target of an epic takeover battle in 1993-1994..." Paramount had originally announced a merger with Viacom. But QVC Network intervened, making a hostile run on the company. During the ensuing bidding war, Viacom strengthened its strategic position by merging with Blockbuster Entertainment, and was finally, five months after the opening shots, able to win the takeover battle.

The quotation is from Niccolo Machiavelli, *The Prince*, A New Translation with an Introduction by Harvey C. Mansfield, Jr., The University of Chicago Press, Chicago and London 1985, pp. 58. Reproduced with permission.

"As the reader will understand, I am not sure that current law..." My thinking on these matters has been influenced by Ouchi's writings on the importance to society of "leaky" intellectual capital, see e.g. W.G.

Ouchi, "Intellectual property and cooperative R&D ventures," in ref. [10]. Insider trading in the futures market and in the bond market is not illegal. In the stock market, prosecution of insider trading in the U.S. has been going on for more than 30 years but has only recently gathered momentum. Some legal authorities point out that insider trading typically involves breaches of trust that could best be dealt with under common law. Corporate executives are permitted to exercise stock purchasing options and to sell stock thus acquired at their leisure. But such transactions by top executives have to be filed with the Securities and Exchange Commission.

"As a matter of fact, many U.S. corporations spend great effort ..." E.M. Remolona, "Understanding International Differences in Leverage Trends," *Federal Reserve Bank of New York Quarterly Review*, Spring 1990, pp. 31-42.

"To illustrate how mergers and acquisitions are carried out ..." B. Burrough and J. Helyar, *Barbarians at the Gate: the Fall of RJR Nabisco*, Harper and Row, New York 1990.

"The mathematical theory that we need is game theory..." J.von Neumann and O. Morgenstern, *Theory of Games and Economic Behavior*, Princeton University Press, Princeton 1944.

Chapter 7

The Long Waves: Evolution or Chaos?

An intriguing aspect of the advance of knowledge in theoretical physics is the link between microcosmos -- the atom and its constituents -- and macrocosmos -- the nebulosae, distant Milky Ways, and black holes. Apparently, knowledge about elementary particles such as protons and bosons and muons can be directly translated into knowledge about the "big bang," the birth of the universe, and its possible future.

There are similar intriguing relations between the microeconomics of nascent technologies on the one hand, and the boom periods of the macroeconomy on the other. The microeconomics of the steam engine is also the macroeconomics of the industrial revolution. Look into the debits and credits of the operation of an oil well or the manufacture of a motor car, and you will have understood the powerful economic upswing that occurred in the Western world during the first part of this century. Try to understand the economics of the semiconductor -- that tiny wafer that you can place on the nail of your index finger -- and you will fathom the roaring advance of the capitalist world during the last quarter of the century sweeping away state socialism in its path, and the birth of the electronic empires of the Far East.

The study of microeconomic gestation, and the analysis of the evolution of economic macrocosmos both belong to the outer fringes of accepted economic knowledge. The conception and the commercialization of new technology is still little understood by economists and psychologists and management specialists. The "long waves" of business cycles are still poorly documented, and questioned by many. And yet, there is a common element in such diverse emergent structures. The common element is creativity itself, and the formal representation of dynamic emergent processes as nonlinear mathematics.

The Long Swings

The long waves in economics are tied to the name of the Russian economist N.D. Kondratieff. Writing in the early 1920s, Kondratieff found two complete such cycles, the first one covering approximately the time period 1790 - 1850, and the second the time period 1850 - 1900. During the upswing a large number of discoveries and inventions made earlier were applied commercially. There was expanded gold production, absorption of new geographical areas into the world economy, rising commodity prices, and often war. The downswings were accompanied by agricultural depressions and falling prices. By Kondratieff's reckoning, he was himself experiencing the upswing into the third such long wave.

Stalin did not like Kondratieff's views -- he preferred the Marxian concept of an imminent collapse of the capitalist system. Kondratieff was sent to a prison camp in Siberia where he perished in 1930.

Joseph Schumpeter set out to provide a theory of long waves by linking them to his ideas of economic innovations in industry. He believed that he could discern a pattern of clustering of innovations during the upswing of the long cycle. He saw cotton and textiles as a "leading sector" propelling Western economies into the upswing of the first Kondratieff, railroads, steam, and steel as the leading sectors of the second Kondratieff, and electricity, industrial chemistry, and the internal combustion engine as the leading sectors of the third one.

The Nobel laureate Simon Kuznets (1901-85) collected data on the secular movements of outputs and prices for wheat, cotton, coal, copper, steel, cement and many other agri-cultural products and raw materials. He documented the long cycles in output, and he showed that periods of accelerating growth go together with falling prices. There are increasing returns to scale during the upswing.

Among modern economists, Walt Rostow has investigated the Kondratieff wave. Writing in the late 1970s, Rostow could draw upon a wealth of new data. The third long wave (presumably ebbing out during the second world war) was then history, and a new wave of growth seemed to be emerging spearheaded by developments in plastics, synthetic textiles, air transportation, mass consumption, and tourism. Was this the fourth Kondratieff in the making? Rostow found that question difficult to answer as he pondered the oil crises of the 1970s and the world-wide recession that followed in the heels of the first oil crisis. Or had the fourth Kondratieff perhaps already peaked, now turning into its downswing?

Rostow pointed at the need to explain the sequence of major inventions and innovations -- the leading-sector complexes -- as well as the incremental improvements in productivity embraced under the case of increasing returns. Economists must explain technology. They cannot leave that to engineers or historians. Technology must be treated as an endogenous variable of the economic system.

Downswings pose another set of problems. Downswings breed doomsday sayers. During the Great Depression in the 1930s many economists came to believe that the modern industrialized economy is inherently headed for stagnation. The 1970s brought a new crop of pessimists. New words were coined reflecting this outlook: stagflation (the occurrence of economic stagnation and inflation at the same time), and Euro-schlerosis (the calcification of the economic arteries of Europe).

In 1972, Dennis Meadows and his collegues published a report entitled *Limits to Growth* which came to have considerable influence. The study was sponsored by the Club of Rome. Analyzing growth trends in world population, industrialization, pollution, food production, and resource depletion, the authors developed computer models that delivered apocalyptic forecasts about our economic future.

The mathematical technique used was computer simulation. The designers of a new airplane can build a small prototype of a new aircraft and test it in a windtunnel. They can also build a mathematical representation of the new aircraft in a computer and test it mathematically under various simulated scenarios. Jay Forrester of the Massachusetts Institute of Technology had conceived of the idea of building a similar computer laboratory for the entire world economy. He had developed a simulation language for computers that he called Dynamo. Following in his footsteps, the authors of *Limits to Growth* collected observations on key economic variables from 1900 to 1970. They built a mathematical model that fitted the observations, including leads and lags. Assuming that no change would occur in the physical, economic, or social relationships that had been observed, the authors thereupon let the computer generate forecasts for the time span 1970 - 2100. In a standard scenario, there would be a global pollution "explosion" toward the mid twentyfirst century. Also, an eroding resource base would lead to dramatically lower industrial output, falling living standards, and eventually, to starvation and a declining world population.

It is doubtful whether *Limits to Growth* should be accepted as serious scientific inquiry. There are no prices in this model. As resources are depleted, no scarcity develops in the sense that prices rise. As a consequence, there is no substitution toward the use of alternative resources. Nor is there any economic theoretical understanding behind the construction of the simulation model. The equations of the model are selected mechanically, to fit the observed data. One important insight in the book, however, redeems many of these deficiencies: pollution, and the environment, should also be treated by economists as endogenous. It is not permissible to treat these matters as known and "given." They are part and parcel of the problem that the economist need to face.

Reversible and Irreversible Time

At this point I should say a few words about the conception of time in economics. Earlier economists tended to look with great envy at their colleagues in natural sciences. The Newtonian world was well ordered and obeyed simple mathematical rules. Economists found it natural to believe that there must exist similar simple mathematical truths in their world waiting to be unlocked. One recurring mathematical regularity was the sinus curve -- the regular wave motions encountered in optics and in the study of sound. The ups and downs of the business world were well known by everybody. What would be more natural than the belief that there exists some underlying basic mathematics that produces such regularity and that the vagaries of the business cycle actually are governed by an underlying strictly recurring cyclic movement?

The Newtonian world was reversible. The physical laws governing a planet orbiting the sun would be precisely the same if the planet were to reverse direction and orbit the other way around. The Newtonian world is like an intricate mechanical clockwork. Turning a few ratchets inside the mechanism upside down, the clock would start ticking backwards instead. The hands would move counter-clockwise. If the time is eight o'clock in the morning right now, one hour later it would be seven o'clock.

One consequence of the great revolutions that have occurred in science during the last one hundred years is the explicit introduction of time in the physical world system. This dramatic change occurred first in thermodynamics, which deals with heat engines and energy. The second law of thermodynamics states that the entropy in a closed system continuously and irrevocably increases toward a maximum. Entropy is a measure of the amount of unavailable energy in a given thermodynamic system. Energy itself cannot be destroyed, but the second law states that energy inexorably moves from available to unavailable states. For example, if we burn a piece of coal, the energy remains but is transformed into sulfur dioxide, carbon dioxide, and other gases that then spread out into space. The available energy has been transformed into unavailable energy. It has become dissipated. Deposits of coal, oil, natural gas and other fossil fuels are nonreplenishable. The extraction of such deposits is an irreversible process. First the more readily accessible strata of the deposits are exhausted. As scarcity develops, and prices rise, it pays to tap deeper and more difficult deposits, such as deepsea drilling, infield drilling, and shale oil.

Following works by Maxwell, Einstein, and others, an explicit time dimension was introduced in the universe of modern physics. The arrow of time moves from the "big bang" to the present. The history of the world is directed in time. But physicists and cosmologists are as yet uncertain whether the universe will forever continue expanding in space-time, or whether it will eventually recollapse in a final big crunch.

Economists, too, have begun to understand that the processes they study are inherently irreversible. The world changes and we change with it. We cannot go back to the economic conditions ruling in the 1920s or even in the 1960s. New technology is developed. Consumer preferences evolve. Tropical rainforests are razed. The stratosphere becomes polluted with greenhouse gases.

Understandably, many people have come to feel that it is necessary to slow down the rapid depletion of the world's nonrenewable resources. The gross national product of any country, as conventionally measured, has a large energy content. "Growth" in the sense of increasing GNP requires more energy. To Nicholas Georgescu-Roegen and others there seemed to be just one conclusion to be drawn: economic growth itself has to be halted. And so the idea of ZPG = zero per cent growth was born. Georgescu-Roegen went even further. A French 1979 translation of some of his writings was entitled *Demain la Decroissance* -- tomorrow negative growth.

In the 1980s, the second law of thermodynamics gained renewed actuality. It is bad enough that the industrial world is fast depleting the stock of usable energy resident in fossil fuels. By the middle of the decade, the realization grew that the concurrent accumulation of unusable energy poses grave dangers, too. The burning of fossil fuels causes carbon dioxide in the upper atmosphere to increase, blocking the release of heat from the planet. Landfills release methane, another "greenhouse" gas. Furthermore, coal burning produces sulfur dioxide and nitrogen oxide. When released into the atmosphere, these gases produce acid rain. Jeremy Rifkin, in his book *Entropy: Into the Greenhouse World* (dedicated to Georgescu-Roegen, "prophet and teacher") envisions a future greenhouse world where New York City is lined with palm trees, massive dikes have been built around Manhattan to hold back the rising seawater, Bangladesh has been flooded by torrential rains and rising flood waters, and desertification claims large sections of central Europe and the American midwest.

The Ebbs and Tides of Basic Innovations

Let me return to the main thread of my story -- the long-term evolution of the U.S. economy. Are there only famines and floods ahead? Is there no silver lining to the future? In the darkness of the stagnating 1970s, a few economists still believed that the capitalist system would rebound. One of them was Gerhard Mensch. He set out to investigate much more carefully than had ever been done before, the time lag between advance in basic knowledge (invention) and the commercialization of such progress (innovation). His conclusion: During prolonged spells of economic stagnation there accumulates a backlog of unexploited technological and commercial opportunities. Eventually the potential of these opportunities will break the stalemate in technology. As the flow of basic innovation swells again, the depression will be overcome.

Mensch has recorded the time span from invention to innovation for more than one hundred basic innovations undertaken in the nineteenth and the twentieth century. Here are two examples out of one hundred:

Chester Carlson (1906 - 1968) succeeded in making the first xerographic copy in 1938. He used an electrostatic method rather than the cumbersome photographic and chemical copying processes of his day. His first patent was dated 1940. For the next four years he tried unsuccessfully to interest someone in developing and marketing his invention; more than twenty companies turned him down. Finally, in 1947, the Haloid Company, Rochester N.Y. (later the Xerox Corporation) acquired the commercial rights and in 1958 Xerox introduced its first office copier. Twenty years from invention to commercialization.

Edwin Armstrong (1890-1954), while still in college at Columbia University, invented the regenerative or feedback circuit. Earlier, radiosignals were only faint whispers barely audible through tight earphones. The new device amplified the signals so that they could be heard through a loudspeaker. But Armstrong's greatest invention was yet to come: in 1933 he invented FM broadcasting. FM (= frequency modulation) modulates the frequency (number of waves per second) rather than the amplitude or power of radio waves. It was an entirely new way of scrambling the sound waves in the transmitter and unscrambling them again in the receiver. It eliminates static. The era of high-fidelity broadcasting had been ushered in. But Armstrong had to fight an uphill battle to get his invention commercialized. The large radio corporations wanted to protect their own systems (short wave and medium wave). They fought him all the way. Exhausted by interminable patent suits and despairing, he finally, in 1954, took his own life. Today FM is the preferred system in radio and the required sound system in television. More than 25 years from invention to commercialization.

Mensch reasons that the rate of basic inventions is a function of the evolution of each particular field of research. Scientific insight, too, is subject to the dynamics of creation, unfolding, maturing, and decline. As long as the frontier of knowledge remains fertile and productive, a steady stream of basic inventions will be made.

Innovations, on the other hand, rest on economic and financial factors. Commericalization of a given technology is profitable when the discounted present value of the expected stream of revenues exceeds the present value of thestream of costs. Innovations tend to cluster during periods of economic boom, when optimism and easy financing prevails. Mensch calculated that the extraordinary progress in basic research in electronics, materials, biology, and medicine during the second half of the twentieth century (to name a few of the most fertile areas) would eventually prompt a prolonged economic boom in the

U.S. Writing in 1975, Mensch predicted that "The surge of innovations will begin in earnest after the year m-s = 1989 - 5 = 1984."

It turned out that he was right. We now know that the bleak 1970s only marked the lull before the avalanche of technological progress that the 1980s would bring: revolutions in microelectronics, new industrial materials, lasers, robots, and biotechnology. The boom economy spurred the commercialization process and shortened the lead time between invention and innovation.

The Diffusion of New Technology and New Products

The theory of diffusion in economics has been pursued most energetically in marketing. Marketing economists try to assess and forecast the degree of acceptance by customers of a new product, and the level of new product sales. Most diffusion models have their roots and analogies in models of epidemics or in biology and ecology. One of the underlying behavioral premises is that new product acceptance is an imitation process, or process of social interdependence. The basic tool of diffusion theory is the diffusion curve. It can be employed to portray the growth of sales of a product over time, such as the annual sales of motor vehicles in the U.S. The curve starts out growing quite rapidly, at an approximately exponential rate. Eventually the rate of increase of annual sales tapers off. The market becomes saturated.

An embryo automobile industry had begun in the U.S. already in the late 1890s. In 1901, Oldsmobile turned out some 1,000 one-cylinder cars, and production doubled in 1902. By 1905, the company was producing more than 5,000 cars per year. The Model T Ford had a four-cylinder engine. Ford sold some 10,000 Model T cars in 1908, 70,000 in 1911, 150,000 in 1912, and in 1914 some 250,000 cars.

To explain rapid growth, one might first try the mathematical hypothesis that the rate of growth of sales stays constant over time. This assumption leads to exponential growth. Demographers routinely employ the exponential curve to forecast the impending population explosion in poor countries. They calculate the excess of the birth rate over the death rate. Drawing an exponential curve, they show how the population will zoom off. Before you know it, there is no standing room left. In a similar fashion, the exponential curve applied to a population of motorvehicles would result in a "motor vehicle explosion." Eventually there would be no parking room left on a single highway.

The diffusion curve is obtained by a simple modification of this mathematical reasoning. In the real world, sales cannot grow without limit. There exists some maximal or potential sales volume at which market saturation occurs. Assume no longer that the rate of growth of sales is constant, but rather that it is proportional to the potential sales volume still left unsatisfied. The solution is the diffusion curve. It starts out growing at an exponential rate but eventually flattens out approaching a saturation level. The curve has a characteristic S-shape. See Figure 1 overleaf.

How does the diffusion curve square with the product "life cycles" that I have discussed earlier in this book? The answer is that the diffusion curve is drawn for an entire category of consumer products. The life cycle refers to a particular product design. For instance, drawing a diffusion curve for the telephone, we can imagine a series of individual product cycles, such as the life cycle for the swivel telephone, the push-button handset, and the cellular telephone. Those life cycles have been indicated with stippled lines in the diagram. During the upswing phase of the diffusion curve, new product cycles rapidly enter the market. During the maturing phase, they arrive at a slower pace.

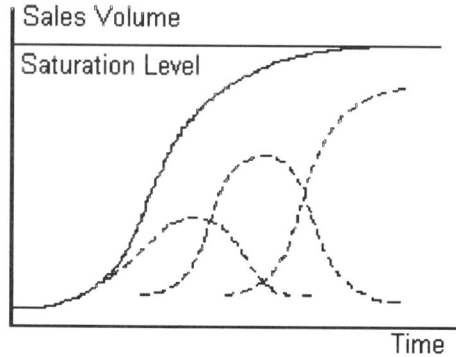

Figure 1. Stylized diffusion curve. Stippled curves indicate individual design life cycles.

Early economists fitted diffusion curves to the sales statistics for many products of the industrial age. Calculations for the automobile industry made by Davis in 1940 indicated that the curve grew exponentially during the early years, reaching 200,000 cars in 1922. But at this point the statistics tapered off, appearing to approach -- but not reach -- a saturation level of a little more than 300,000 cars a month. Although automobile sales remained brisk the industry was, in the late 1930s, no longer a growth industry. Today, the U.S. motor vehicle industry turns out close to to 1 million cars and light trucks a month (the figure includes some 100,000 Japanese cars built in this country). Look at Figure 1 again! New waves of motor car product cycles continued to carry the diffusion curve upwards.

The most well-known model of innovation diffusion is due to Frank Bass. Bass divides all potential customers of a new product into two classes: innovators and imitators. Sales to innovators grow at their briskest pace when the product is new, but slower when adoption becomes more common. Sales to imitators follow the diffusion curve. The original Bass paper (1969) made a forecast of the timing and magnitude of the peak of color television set sales. At the time the forecast was made, only three years of data were available. The forecast proved to be extremely accurate. In a recent study of the Bass diffusion model, empirical results were reported for room air conditioners and clothes dryers (examples of consumer durables), for ultrasound and mammography (medical innovations) and for the study of foreign language and accelerated programs (educational innovations). In all cases the statistical fit was quite remarkable.

The general idea that evolution takes the form of a succession of generations of "species," be it economic species or biological species, is as old as the concept of evolution itself. In the marketplace, a novel and attractive product will capture market share that would otherwise have belonged to competitors selling outmoded product designs. The same phenomenon occurs in the biological world when dandelions and crabgrass take over a lawn. The capture occurs as a crowding out, the taking of potential living space.

The mathematics of successive generations of species was developed in the 1930s by the Italian mathematician Vito Volterra. His classical treatise is entitled *Lecons sur la Theorie Mathematique de la Lutte pour la Vie*, which translates as Lessons on the Mathematical Theory of the Struggle for Life. Here one finds the origin of the modern nonlinear mathematical theory of evolution. But Volterra had to conduct his pioneering research with a severe handicap: he had no electronic computer. He never became aware of the amazing numerical possibilities to which I now turn.

The Possibility of Chaos

As it so happens, the diffusion curve can be used to illustrate the phenomenon of chaos. In 1975, the mathematician J. Yorke was playing around with the curve, looking at what happens when one modifies its mathematical structure, replacing continuous time by "discrete time." A TV broadcast of a football game is in continuous time. An eight millimeter home movie of the same game is in discrete time. Time is cut up into a series of still segments. Movements are "jerky," but our eyes cannot see the difference if the movie is run at the right speed. The step from the continuous version to the discrete version may seem like a small modification.

Under suitable numerical conditions, the discrete time path of the diffusion curve does indeed converge toward the continuous path. It may temporarily overshoot the saturation level, but eventually zooms in toward the equilibrium value. Increasing a numerical coefficient regulating the "boom-and-bustiness" of the dynamic course of events, Yorke made a startling discovery. The curve became so jittery that it "bifurcated" or split into two. The annual sales volume ended up oscillating between two alternative values, one greater saturation level and another smaller saturation level.

If the coefficient of boom- and-bustiness is increased even further, beyond the value of three, the discrete path breaks up into chaos. It looks unorganized and random. A wild dance. And yet, it has been generated from a very simple and deterministic mathematical rule. Yorke titled his paper "Period Three Implies Chaos." It was published in a mathematical journal in 1975. A new science had been born.

To see how bifurcations and chaos can occur as one product cycle replaces another, please refer to Figure 2 below. This diagram plots the results from a simple computer simulation, assuming that there are three consecutive product cycles. Sales of the first generation of the product take off almost immediately, reaching a peak level of 0.117 in month 10, and then slowly taper off. Already in month 23, sales of the first generation of the product have virtually vanished. The second generation comes on line a little bit later, reaching a peak of 0.309 in month 12. Sales of the third generation of the product, finally, appear in

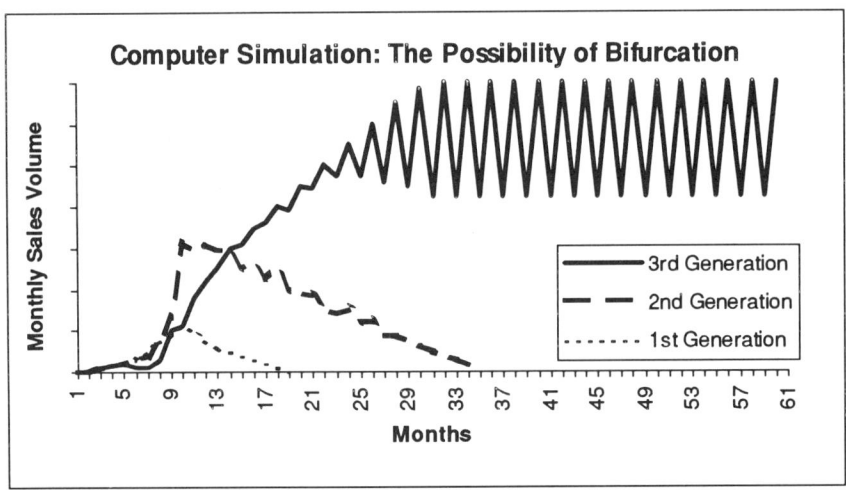

Figure 2. Sales statistics for three consecutive product generations. Computer simulation. The third generation product starts bifurcating in month 19.

month 5 and grow steadily at first, hitting 0.403 in month 18. Then something unexpected happens: the monthly sales statitistics starts zigzagging, breaking up into two different paths, one upper path and one lower path. The sales curve bifurcates.

The case of chaos is illustrated in Figure 3. This time the numerical coefficient measuring the boom-and-bustiness of the third generation of the product has been increased slightly, enough to tip the dynamics of the system over the edge of chaos. Sales of the two first generations follow the same kind of life cycles as before, with the total sales volume slightly smaller. Sales of the third generation come on line in month 5 ; from there on sales grow rapidly reaching 0.558 in month 12. But the subsequent statistics should be a nightmare for any market analyst. In one month a glorious upturn; in another, the pits. Notice the lull during the months 20-25. But, as it turns out, this brief interlude of stability is just the the calm before the storm.

Bifurcation indicates that the sales statistics is poised at the "edge of chaos." It is a balancing act between order and full-blown chaos. It is a fleeting moment before the system takes the plunge into disorder. It is the sudden appearance of alternative futures, a forking point along the road of technological advance, such as two competing product designs.

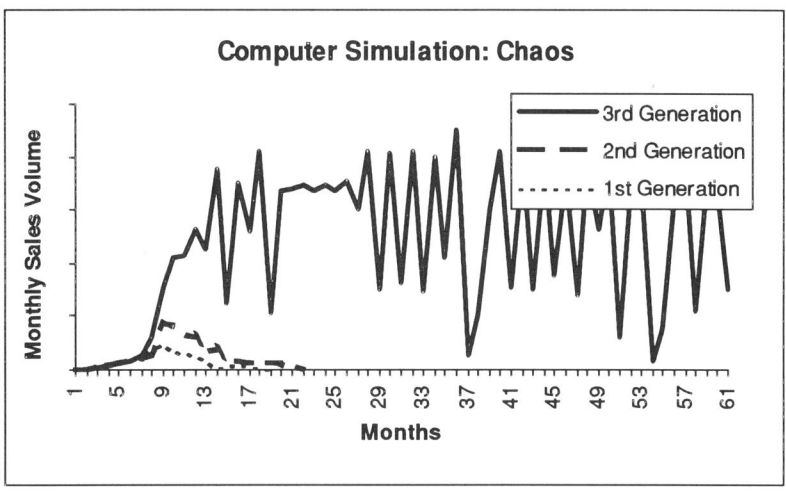

Figure 3. Sales statistics for three consecutive product generations. Computer simulation. Sales of the third generation product turn chaotic in months 11-13.

Is chaos the same thing as randomness? In the computer simulation, chaos is produced by a few simple (nonlinear) mathematical equations. The sales curve in Figure 3 is determined, point by point, by an exact mathematical formula. In the lingo of mathematical statistics, it is "deterministic." Randomness, on the other hand, is the result of probability drawings, like the random throw of dice, or the spinning of a roulette wheel. Chaos *looks* random. But there is order hidden behind the apparent whimsicalness.

In a zoological experiment, a population of ants is let onto a marble floor strewn evenly with decomposing leaves and sticks. They get to work, carrying and dragging. In a couple of hours, the start-ups of several competing ant heaps can be discerned. After several days, one heap wins out and all ants are busy coordinating their efforts, building the same heap. Order has emerged, out of what initially seemed to be chaos.

> **Ilya Prigogine -- Scientist and Humanist**
>
> When it bestowed its prize for outstanding contributions to European culture on Ilya Prigogine, the Italian National Committee for the Umberto Biancamno Prize cited him for furthering the "conceptional evolution of scientific understanding, which has called so many minds in both the old and the new Europe to take new stands in creative and innovative thought."
>
> Ilya Prigogine was awarded the 1977 Nobel Prize in chemistry for his contributions to nonequilibrium thermodynamics, particularly for the theory of dissipative structures. He is the director of the International Institute of Physics and Chemistry (Solvay Institutes) in Brussels, Belgium and he is a Regental Professor of Physics at the University of Texas at Austin. Recently, he received the title of Commandeur de la Legion d'Honneur. The king of Belgium has made him a viscount.
>
> Prigogine has emerged as the proponent of an evolutionary conception of the world, replacing the Newtonian mechanistic paradigm. He has shown that the behavior of matter under non-equilibrium conditions can be radically different from its behavior at, or near equilibrium. This difference introduces multiple choices, self-organization, and complex dynamics.
>
> Close to equilibrium, the description of the temporal evolution of a system can be expressed by linear equations. Far from equilibrium, one deals with nonlinear equations, which may lead to bifurcations and to the spontaneous appearance and evolution of organized states of matter, so called dissipative structures. As an example of a dissipative structure, consider a pan of liquid heated from below. When the temperature is low, heat passes through the liquid by conduction. As the heating is intensified, regular convection cells appear spontaneously. The liquid boils. Energy is transferred from thermal motion to convection currents. The boiling dissipative structure is radically different from the equilibrium structure of the liquid. It can be maintained in far-from-equilibrium conditions only through a sufficient flow of energy. Yet it represents order.
>
> Within the context of the new paradigm, the world is no longer seen as a vast automaton. Instead, it is subject to self-organization and evolution. Prigogine views energy dissipation as the driving force of evolution. He notes that, despite the increase in organization and complexity of living systems, there has been an acceleration of biological evolution in the course of time. Each new step increasing the functional organization has in itself the germs for further evolution. For instance, mathematical relations describing the evolution of thermodynamical systems can be adapted to understand the notion of survival of the fittest in predator - prey ecosystems. The prey evolves so as to exploit the available resources more efficiently, and to avoid capture by the predator. The predator, on the other hand, evolves so as to increase the frequency of capture of the prey and to decrease its death rate. The ratio of the biomass of predator to prey increases slowly with evolution.

Sometimes one might feel that the city of New York is a big ant heap. The sidewalks are filled with people rushing to and fro. Where are they all headed? Is there some system to this madness or are these just human antics, random

The Long Waves: Evolution or Chaos? 105

wanderings? The point is that it may be difficult to say precisely what "random" is. A phenomenon may look random to the observer, and yet have perfect purpose.

By the nature of things, the future will always seem more chaotic then the past. During the bleakest hours of the Second World War, many people came to doubt the future of the entire Western civilization. What they saw, was just the abyss. In retrospect, we know that the dictatorships of Hitler and Mussolini and Tojo were to fall, eventually to be replaced by stable democracies. In retrospect we can see the pattern, the cyclical fall and rise. We discern the order behind the chaos.

But prospectively, while the process is still an ongoing one, we do not yet have all the answers. There is no doubt a pattern to the collapse and eventual reemergence of Yugoslavia or Somalia, but we cannot see it now. There is also no doubt a pattern to the shrinking industrial base and the loss of manufacturing jobs in the U.S., and the eventual resolution of the economic and human problems that follow in the wake of such structural change. But that pattern may be elusive to city officials and to the general public that are wrestling with increasing unemployment, rising crime, and a groundswell of juvenile violence. What we see right now is just the chaos, the chaos in the homes and the chaos in the streets. The breakdown of order.

Deepgoing change of nonlinear systems, be those systems physical systems like the climate or economic systems like the U.S. computer industry, will to the contemporary observerer almost invariable look disorderly. One region of the country will be drenched in floods, another will suffer the longest spell of draught of the century (this all happened during the summer of 1993). Yet there may be order: global warming. The largest and most successful computer companies of the nation are brought to the brink of bankruptcy. Yet, there is order: the advent of new microprocessor technology.

The lesson seems to be that an existing system can only change when it is brought to the limits of what it can endure. At that point of extreme strain, the old and the obsolete will be shed . The new will rapidly take over.

Turbulence and Shake-Out in U.S. Industry

It is one thing to generate bifurcations and chaos on a computer. But do these concepts have any relevance in the real world?

Remember that each single product design is really a hierarchy of technologies and subassemblies. As product development brings forth new generations of a product, the underlying logistics chain adjusts. New links are set up in the industry network; others are falling by the wayside. There is a rerouting of the flow of goods and services throughout the economy. I have earlier discussed such change under the headings of budding and cross-fertilization. Budding is the upgrading of existing technologies and existing products. Cross-fertilization is the leap into the unknown, the spark that joins together separate technologies into new combinations.

The dynamics of a sequence of product cycles therefore translates into a dynamics of rerouting and change in the underlying logistics network. An orderly sequence of product cycles will correspond to budding and orderly evolution of the complexity of the network. Bifurcations and chaos mirror the confusion of cross-fertilization, the rapid growth of small startup companies employing novel technology, and the gradual erosion of companies using outdated technology.

These notions seems to have a peculiar application to current events on the industrial scene in the U.S. The advance of high technology such as computer technology and biotech involves the formation of ever higher and more differentiated production hierarchies. Rapid advances in information technology and the globalization of trade, commerce, and finance is leading to new managerial hierarchies, including the outsourcing of manufacturing to subcontractors domestically or abroad, and the breakup of large monolithic management structures into smaller and more versatile units. Are we now, in the early1990s, witnessing a "phase transition" in the U.S. industry, a shake-out at the edge of chaos?

One word comes to mind: polarization. That is, the industry seems to be breaking up into two distinct camps. The first camp consists of small and young companies, recent startups located at the beginning of their life cycles. They have increasing returns to scale and, consequently, face a financial incentive to grow. They concentrate their activities around one or just a few products. Further growth depends upon their ability to launch in a timely fashion new successor generations of the original product.

The other camp consists of established and mature companies that typically have achieved considerable size and that manufacture and market a wide range of products. They have decreasing returns to scale, and many of them display signs of acute distress (falling sales, negative profits). The aging corporations face a formidable task: marketing a wide range of products, they need to upgrade or renew a large number of entries in their product line every year. Many large companies have difficulties staying cost-effective while they accomplish such a broad-based R&D program. They are eclipsed and outperformed by swarms of small startup companies.

The picture of U.S. industry, then, is one of turbulence. There is no apparent "equilibrium." The turbulence is a confrontation between individual products and technologies located at different phases of their life cycles. It is also a confrontation between individual corporations, some being more effective than others in their product cycle management: some companies are riding on the upswing of their cycles, others struggling as their products are reaching maturity and eventual obsolescence.

Bibliographic Notes

Mathematical formulations of dynamic economics go back at least to the Frenchman A. Cournot (in his *Recherches sur les Principes Mathematiques de la Theorie des Richesses*, Hachette, Paris 1838, English ed. MacMillan, New York 1927). Virtually all dynamic economics before 1960 was linear dynamics. The time path of a linear dynamic system can exhibit only three patterns: convergence, divergence, and oscillation. The well-known dynamic systems of the "cobweb" (depicting the development over time of the price formation in a single market, see A. Hanau, "Die Prognoze der Schweinepreise," *Vierteljahrshefte zur Konjunkturforschung*, Vol. 2, 1927) and the interaction of the multiplier and the accelerator (P.A. Samuelson, "Interactions between the Multiplier Analysis and the Principles of Acceleration," *Review of Economic Statistics*, May 1939, pp. 75-78) are linear. With suitable numerical values specified for the parameters of the system, they will be convergent.

Nonlinear dynamic systems typically exhibit more complex patterns, including crises, bifurcations, and other chaotic phenomena. The last 10 to 15 years of economic research has picked up the standard models of periodic equilibrium and showed that chaotic fluctuations are consistent with them. The role of bifurcation theory is to demonstrate how and when periodicity gives way to chaos. It is now widely accepted that chaotic dynamic patterns may theoretically arise in a well-functioning market economy. (See e.g. J.M.Grandmont, "On Endogenous Competitive Business Cycles," *Econometrica*, Vol.53, 1985, pp. 995-1045.)

In particular, chaos theory seems to be helpful to understand the dramatic changes or "paradigm shifts" that can transform an entire industry (the introduction of color TV, jet propulsion in the commercial aircraft industry, the arrival of the personal computer). In the language of nonlinear dynamic systems, such a shift is referred to as a "phase transition." It can be generated mathematically by changing the numerical value of a system parameter.

References

"...the Russian economist N.D. Kondratieff" An excellent and very readable survey of technological history and the long waves is given in R.U. Ayres, "Technological Transformations and Long Waves. Part I-II," *Technological Forecasting and Social Change*, Vol. 37, 1990, pp. 1-37 and pp. 111-137.

"Among modern economists, Walt Rostow has..." See W.W. Rostow, *The World Economy: History and Prospect*, University of Texas Press, Austin 1978 and W.W. Rostow, "The Fourth Industrial Revolution and American Society: Some Reflections on the Past for the Future," in ref. [8].

"In 1972, Dennis Meadows and his colleagues published..." D.H. Meadows, D.L. Meadows, J. Randers, and W.W. Behrens III, *The Limits to Growth*, Universal Books, New York 1972.

"Following works by Maxwell, Einstein, and others, an explicit time dimension was introduced..." I have found R. Penrose, *The Emperor's New Mind: Concerning Computers, Minds, and the Laws of Physics*, Oxford University Press, Oxford 1989 particularly instructive.

"To Nicholas Georgescu-Roegen ... there seemed to be just one conclusion" N. Georgescu-Roegen, *Energy and Economic Myths*, Pergamon Press Inc., New York 1976. The French translation is *Demain la Decroissance*, Pierre-Marcel Favre, Paris, 1979.

"Jeremy Rifkin ..." J. Rifkin with Ted Howard, *Entropy: Into the Greenhouse World*, Bantam Books New York 1989, originally published as *Entropy: A New World View*, Viking Press 1980.

"One of them was Gerhard Mensch. He set out to investigate..." G. Mensch, *Das Technologische Patt*, Umschau Verlag, Frankfurt 1975, English language edition *Stalemate in Technology: Innovations*

Overcome the Depression, Ballinger, Cambridge, Mass., 1979. The quotation in the text is from *ibid.*, p. 197.

Figure 1. Stylized logistic curve. I should point out that the curves that I have drawn here are almost identical to a diagram on p. 457 in G. Nicolis and I. Prigogine, *Self-Organization in Nonequilibrium Systems: From Dissipative Structures to Order through Fluctuations*, Wiley, New York 1977, illustrating the evolution of ecosystems.

"Early economists fitted logistics curves ..." H.T. Davis, *The Theory of Econometrics*, The Principia Press, Bloomington, Indiana 1941. For extensive use of the logistics curve to forecast new technology, see H.B. Stewart, *Recollecting the Future: A View of Business, Technology and Innovation in the Next 30 Years*, Dow Jones-Irwin, Homewood, Ill., 1989.

"The original Bass paper (1969) made a forecast of ..." F.M. Bass, "A New Product Growth Model for Consumer Durables," *Management Science*, January 1969 pp. 215-227. See also F.M. Bass, "TheAdoption of a Marketing Model, Comments and Observations" in Ref. [15].

"In a recent study of the Bass diffusion model..." see V. Mahajan, C.H.Mason and V. Srinivasan, "An Evaluation of Estimation Procedures for New Product Diffusion Models" in Ref. [15].

"The Possibility of Chaos " For a non-technical account of the theory of chaos, see J. Gleick, *Chaos: Making a New Sience*, Viking Penguin Inc., New York, N.Y. 1987. See also M.M. Waldrop, *Complexity: The Emerging Science at the Edge of Order and Chaos*, Simon & Schuster, New York 1993, and R. Lewin, *Complexity: Life at the Edge of Chaos*, Macmillan, New York 1993

"In 1975, the mathematician J. Yorke started playing around with ..."J.Yorke and T-Y. Li, "Period Three Implies Chaos," *American Mathematical Monthly*, Vol.82, 1975, pp. 985-992.

Figure 2. The computer simulation illustrated here was set up as follows. Let X be the sales of the first generation, Y the sales of the second generation, and Z the sales of the third generation product. The initial values were put at X= 0.001, Y=0.0001, Z=0.00001. Furthermore, sales in any subsequent month (denoted +1) was assumed to be related to sales in the previous month by the equations $X_{+1} = 2X(1-X-Y-Z) - 0.1X$, $Y_{+1} = 3Y(1-X-Y-Z) - 0.25Y$, $Z_{+1} = 3.75Z(1-X-Y-Z) - 0.4Z$.

Figure 3. The numerical assumptions are the same as for Figure 2, but the sales for the third generation product have this time been set at $Z_{+1} = 4.35Z(1-X-Y-Z) - 0.4Z$.

"In a zoological experiment, a population of ants..." Experiments of this nature have been reported in E.O. Wilson, *The Insect Societies*, Harvard University Press, Cambridge, Mass. 1971.

Box on Ilya Prigogine. The text is based on G. Nicolis and I. Prigogine, *Self-Organization in Nonequilibrium Systems: From Dissipative Structures to Order through Fluctuations*, Wiley, New York 1977. See also I. Prigogine and Stengers, *Order out of Chaos, Man's New Dialogue with Nature*, Bantam Books, Toronto 1984.

CHAPTER 8

Genetic Selection and Biotechnology

The strides of technological progress in engineering and information systems have been astounding. But the advances in agriculture, livestock production, and the fisheries are also impressive. Not only is man a master of machines, but he is on his way to engineer and control "nature" as well.

Tinkering with nature, however, can pose great potential danger. The ecological balance may become upset. As we shall see, the technological advance in the life sciences demonstrates not only the ingenuity of the human creative mind but also its limits.

Early Technology on the Farm

In an important sense, the history of civilization is the history of technological progress in agriculture and livestock production. The first step in this evolution was taken when man left the stage of a nomad hunter-gatherer and settled in villages, cultivating the land. The technological breakthrough that made this possible was the domestication of wild wheat. The so-called Emmerwheat which has 28 chromosomes, arose through the hybridization of wild wheat and goat grass which each have 14 chromosomes. A further hybridization between Emmer and another natural goat grass produced bread wheat which has 42 chromosomes. The ear of bread wheat is plump and heavy and the ear is too tight to breakup to allow the seed to spread in the wind. From that moment on in history, the propagation of wheat was dependent upon man, just as man was dependent upon wheat.

The civilizations in Mesopotamia and the Nile valley arose as an increasing agricultural surplus made it possible to sustain a class of city dwellers, craftsmen, civil servants, and administrators. The Sumerian city of Ur included storehouse recorders, work foremen, overseers, and harvest supervisors. The land was plowed by teams of oxen, and the grain was harvested with sickles. Wagons had wheels with leather tires. In Egypt, about 3000 BC, there was a ministry of agriculture, a chief of the fields, a master of largesse looking after the livestock, royal domains and temple estates. Irrigation and the waters of the Nile were controlled. The land was tilled with a wooden plow. Crops of barley were cut with sickles. One breed of cattle was kept for meat and another for milk.

The Roman Cato the Censor estimated that a typical olive grove required the following equipment: three large carts, six plows, three yokes, six sets of ox harness, one harrow, manure hampers and baskets, three pack saddles and three pads for the asses. The required manpower included an overseer, housekeeper, laborers, teamsters, a muleteer, swineherd and shepherd.

Already in medieval times, the Netherlands became an agricultural example to all Europe. Dikes were built to protect against floods, and land was reclaimed. Leguminous and root crops were introduced into crop rotation. Town refuse was added to the supplies of animal manure. Dutch cattle was crossed with animals from northern Italy. Flemish horses were known for their size and strength.

Mechanization on the farm arrived around the middle of the nineteenth century. John Deere, an Illinois blacksmith, built in the 1830s a new kind of plow entirely made of steel. Cyrus Hall McCormick demonstrated his new reaper at the Great Exhibition of London in 1951. Steam power was widely adopted on large farms for threshing, shredding and shelling. Super phosphate was patented in 1842, the first synthetic fertilizer.

The technological advances that I have briefly reviewed up to this point are of several kinds. There was a gradual development and accumulation of real capital on the farm, both fixed capital like buildings, wells and terracing and irrigation, and capital in the form of machinery and equipment. There was an improvement in management and in agricultural practice (like crop rotation). There was also something quite notable: development of "nature" itself, like the domestication of wild species of grain and cattle, and cross-breeding and selection to create new and superior varieties.

The Creation of New Varieties and New Species through Hybridization and Selection

In a primitive sense, genetic selection occurs whenever a farmer selects seed from a plant or offspring from an animal that seems to possess superior qualities. Repeated over a span of thousands of generations, the result will be to create an entirely new species. Cross-breeding or hybridization is a much more sophisticated process, but was well understood as a practical matter long before the emergence of the modern science of genetics. Using cross-breeding and selection, the Dutch created many-splendored tulips, and the English their hunting dogs. The U.S. experimenter Luther Burbank developed the Burbank potato and many new varieties of fruits and vegetables, without any formal knowledge of genetic principles.

The theoretical groundwork for the methods of genetic improvement of plants was laid by Mendel (the laws of genetic inheritance) and Darwin (the variation of life species). Mendel was a monk in Brunn, Austria (now Brno, the Czech Republic) who crossed garden peas in his monastery garden and discovered that the characteristics of the offspring was distributed according to precise mathematical laws. One of the first agricultural products created through systematic application of the new genetic understanding was hybrid corn, created through hybridization of several individual kinds of maize and introduced in 1921. In the beginning, hybrid corn represented only a small proportion of all maize grown, but came eventually to change much of American agriculture.

The Santa Gertrudis beef cattle were developed on the King Ranch in Texas by crossbreeding Shorthorns and Brahmans, a heat- and insect-resistant breed from India. The Brangus breed was obtained by crossing Brahman and Angus cattle. -- Artificial insemination was demonstrated by Ilya Ivanov in Moscow in 1919.

Successful hybridization and cross-fertilization does not differ in kind from the sort of technological progress and innovations that we have already met in manufacturing. They push the technology frontier forwards. Sometimes there are large discontinuous jumps. Most of the time, there are only modest improvements. As a result of the technological improvement now indicated, there has been a momentous advance of productivity in U.S. agriculture. Today U.S. farmers and ranchers representing less than 2 % of the country's economically active population, provide the nation with abundant food supplies and a huge export surplus.

The single most important factor limiting crop yields worldwide is soil infertility. Lack of sufficient nitrogen is by far the most widespread. Phosphorus and potassium deficiencies are also common. There thus exists a very considerable potential for increasing crop yields by applying the right amounts of the right kind of fertilizers.

The breakthrough in wheat and rice production occurring in the developing world since the mid-1960s has come to be known as the "green revolution." The green revolution is not just a transfer of high-yield technology from developed countries to peasant farmers in the third world. It was the start of a process of using principles of modern agricultural science to develop technologies appropriate to the conditions of farmers in Indian, Pakistan, the Philippines, China and other third world countries.

The green revolution was spearheaded by the International Rice Research Institute (located in the Philippines), the International Center for Maize and Wheat Improvement (Mexico), and a network of international agricultural research centers. Scientists developed disease-resistant, semidwarf wheat and rice varieties with radically improved yields. Even using traditional methods of cultivation, the new varieties were more efficient in converting sunlight and nutrients into grain. Grown with adequate moisture and soil fertility, they yielded up to four times as much. The new varieties look different from the traditional types; they also require different care in the use of fertilizer, in water management, and in weed control. When the peasant farmers saw the results of the new technologies, they readily adopted them. The combined rice and wheat output of the less developed countries increased by 74 % between 1965 and 1980, with only a 20 % increase in the area planted.

Pesticides and Herbicides

But the advances in agricultural technology are a two-edged sword. The extended use of fertilizers lowers the humus content of the soil and makes it prone to erosion. Further, as pasture lands and forests and woodland areas are being converted to cropland, soil erosion is accelerated, and wildlife habitats and recreational areas are destroyed. Much of the recently cultivated land in the U.S. is marginal and highly susceptible to erosion. Legislation passed in 1985 provides a land set-aside program that pays farmers to return fragile lands to pasture for at least ten years.

Herbicides and pesticides contain toxic elements that may find their way into the food chain eventually or enter the ground water. One of the first synthetic pesticides was DDT, hailed as "the atomic bomb of the insect world." After the second world war, farming in the United States underwent rapid change as farmers turned to the wholesale use of chemical fertilizers, tractors and DDT. But the synthetic pesticides soon showed a darker side. Insects became resistant. DDT began to accumulate to lethal levels in the food chain, killing birds and fish. It appeared in mother's milk. In 1962, Rachel Carson wrote the book *Silent Spring* alerting the general public to the dangers. The Environment Protection Agency (EPA) was created in 1970 and DDT was banned in the United States in 1972.

In order to fight the war against pests and weeds, the chemists need ever larger, stronger, and more varied doses of poison. In 1980 there existed about 35 000 different kinds of products for fighting insects, weeds, and fungi. It is not clear that the chemists are winning. One of the difficult pests that attack cotton is the boll weevil which entered Texas from the south in 1892. By 1921 it had crossed the Mississippi. It will ravage up to half of the harvest on a cotton farm. Pesticides have not helped much. Farmers have learnt to switch to other cash crops. The Texas "blacklands" (between Dallas and San Antonio) used to provide more than a third of the state's cotton; today the figure is less than 10 percent. In the entire nation, cotton receives 26 million pounds of toxaphene alone. This is an insecticide that in the laboratory causes tumors in mice and deforms catfish.

Foresters, utility work crews, rice growers, and ranchers use herbicides to eliminate brush and weeds. In Vietnam, jungles were defoliated with Agent Orange, a mixture containing extraordinary amounts of the same herbicides. The active ingredient is dioxin. Small doses of dioxin cause rhesus monkeys to stop their production of red and white blood cells. Male monkeys stop their production of sperm cells, pregnant female monkeys abort.

Chemical agents are used in agriculture to enhance the appearance or storability of the harvest. Examples are the spraying of apples (the Alar scare), grapes, oranges, tomatoes. The possible toxic effects of such agents is still debated.

Chemicals are also used in livestock production. Chicken feed is laced with antibiotics. Cattle is fed with hormone-treated feed. On January 1, 1989, Europe's Common Market banned the importation of such meat. The U.S. offered to put the health issue before an expert panel. The offer was rejected. Cows injected with the growth hormone bovine somatotropin (bST) produce up to 25 % more milk. But several supermarket chains have refused to carry hormone treated milk.

The problem posed by pesticides and other chemicals is that their true cost does not appear in the profit-and-loss calculation of the individual farmer. His only cost is the cash paid for the pesticide or herbicide. The cost of disease and death is borne by workers, consumers, or even by people who participate neither in the production nor the consumption of the agricultural product. Economists call such costs that appear outside the profit-and-loss statement of a producer "external costs," or "negative externalities." Whenever such costs are present, the conventional profit and loss calculation of a technology gives an erroneous picture of the true cost to society. In the books of the farmer, a pesticide may appear to be profitable. But to society there is a cost to be borne. There is a discrepancy between the profit calculation of the individual producer and for society as a whole.

Economists have spent considerable effort to understand the presence of negative externalities. Pesticides and herbicides are released in the environment. The environment is not "owned" by any one single person or corporation - -it is a common resource of society. There are no individual property rights to environment. Hence, no property owner will claim recompense when a polluter releases toxic substances into the environment. The market price paid by the polluter for discharging the toxic substances is nil. Negative externalities occur also in manufacturing, whenever a corporation releases harmful or undesired substances into the environment. It happens when a glass grinding mill releases powder glass into a stream, when a steel mill releases poisonous fumes into the air, and when toxic waste in the chemical industry is dumped without proper safeguards. The theoretical principles are the same. The market mechanism fails because no property rights are defined to the environment, and therefore no individual owner can claim recompense for the use (and abuse) of the environment.

> ### A Visit To Malcolm Brown's Laboratory.
>
> Malcolm Brown is a professor of botany at the University of Texas at Austin and a senior research fellow at the IC^2 Institute. He is doing pathbreaking research in the biosynthesis of cellulose by bacteria, and the possibilities of industrial production of cellulose by growing bacteria on a glucose substrate. In brief, the idea is to supplement forestry and cotton farming by growing cellulose in laboratories.
>
> As every schoolchild knows, Louis Pasteur invented a method for killing off that whitish fruit mold that tends to form on wine or beer during fermentation. The method is called pasteurization. The fruit mold is formed by a bacterium called *Acetobacter* which has the ability of "synthesizing" cellulose -- manufacturing it, drawing energy from glucose. The fruit mold is pure cellulose.
>
> In a sort of reverse pasteurization, Malcolm Brown grows the *Acetobacter* in his laboratory. He has spent much of his life examining the process. He has studied how the bacterium grows microfibrils from its surface into the culture medium where they interlace, forming twisted ribbons of cellulose. Ribbon size depends upon age of the cell, or the specific strain.
>
> What is cellulose? It is a so-called "homopolymer" or a macromolecule -- actually the most abundant macromolecule on earth. The microfibril is the basic structural unit of cellulose. The fibrils are arranged in specific patterns within plant cell walls. The plant cells are largely dead when the wood or fibers are harvested.
>
> The biosynthesis of cellulose in Malcolm Brown's laboratory opens the possibility of growing cellulose "made to order," in pure form. Using chemical agents which interfere with the crystallization of cellulose, it is possible to control the formation of microfibrils into ribbons. As examples of such cellulose engineering, Brown demonstrates a thin cellulose membrane suitable for making paper money, and a cellulose glove that he has grown to fit his own hand.
>
> Brown pulls out a drawer and demonstrates a file of patent applications relating to the commercial exploitation of his ideas. At one time he figured that it should be possible to grow genetically modified bacterial cellulose in large vats in the desert in West Texas. There should be sunlight enough. Glucose is obtained by a process known as photosynthesis. If the paper and pulp industry in the world could cover its needs for cellulose by growing it on glucose substrate, there would be vast beneficial effects on the environment. Forest tracts in both the North and the South hemisphere could be restored to their original pristine status, offering opportunities for human recreation and for wildlife.
>
> More recently, Brown has gotten an even more revolutionary idea. In 1990, he and his colleagues were able to identify the gene in *Acetobacter* responsible for the production of cellulose and to clone it. Perhaps it would be possible to transfer the gene into the DNA of other organisms, such as seaweed. Consider the possibility that the paper for tomorrow's newspapers and books might be harvested on the high seas!

There is just one solution to this dilemma: society must protect itself. Hence the Environmental Protection Agency (EPA) and the Superfund. The

Comprehensive Environmental Response, Compensation and Liability Act of 1980 (the "superfund") makes those that generate, transport, or store hazardous waste liable for its cleanup, regardless of whether their actions were legal at the time they took place. Individual states have also enacted statutes modeled after the superfund. The EPA along with state agencies is working together to solve three main problems: contaminated ground water, properties with chemical pits, and sites where waste was abandoned. The EPA has established new cleanup technology for such sites. Bioremediation involves the adding of nutrients to contaminated wells in order to break down hazardous chemicals into nontoxic substances. In situ chemical extraction refers to the injection into the soil of acidic or alkaline solutions. The substances are flushed from the subsurface and collected at recovery wells. Vapor stripping involves forcing air through contaminated soils. The vapors rise through the soil and are collected on the surface.

One problem of the pesticide war fought on our farms is that new insect strains emerge that are resistant against pesticides. Such emergence can be seen as a case of Darwinian selection and "survival of the fittest." In nature there is always some genetic variation between individuals belonging to the same species. A pesticide may kill off most individuals, but a few may by chance be endowed with genes that enable them to survive. As these mate, there is a high probability that the same genes will be transferred to the offspring. A new pesticide-resistant generation follows. Since many insects pass through six generations in a growing season, the genetic change of the species through natural selection may occur quite rapidly.

Among the 70 major insect pests that plague corn, the root worm is one of the most difficult to control. It was introduced in Nebraska in 1960 and fanned rapidly out to infest the entire corn belt. It has now become resistant against a once potent poison called heptachlor. Such man-made chains of carbon, chlorine, and hydrogen affect the nervous system.

The confrontation between the chemical engineer trying to design an efficient pesticide, and the evasive action taken by nature in creating new resistant strains, can be seen as a fight between two evolutionary systems: the efforts of man and the efforts of nature. Man designs a spectrum of pesticides with ever more advanced chemical engineering. Nature designs new species that are able to survive. Both systems learn. One generation of chemical engineers learns from the technological mistakes and failures of the preceding generation. One generation of insects learns from the genetic mistakes and failures of the preceding one. The task of managing agricultural policy is thus one of weighing the desire for rapid increase of crop yields against a complex of environmental concerns.

Genetic Engineering on the Farm

Biotechnology was born as man learnt to understand and manipulate the genetic makeup of living organisms -- bacteria, plants, animals, and man himself. Biotechnology is the use of living organisms to deliver new drugs, crops, fertilizers, and pesticides. The breakthrough discovery was the description, by J.D. Watson and F.H.C. Crick in 1953, of the molecular structure of deoxyribonucleic acid (DNA). A gene is a particular sequence of DNA. During sexual reproduction, some genes from one parent are combined with some genes from the other parent to produce offspring containing a new arrangement of the genes. Through genetic engineering in the laboratory (recombinant DNA), it is possible to create new combinations of genes.

For instance, there is a bacterium (*Pseudomonas Syringae*) found on plant leaves that forms a nucleus for ice crystallization. Plants then freeze at a temperature a few degrees above the freezing point. Scientists have been able to locate the particular gene that makes the bacterium an attractive ice nucleus, and to snip out that gene while leaving the rest of the bacterium intact. Spraying such altered bacteria on plants, they will displace the nearly identical ice-forming organism normally found in nature. In 1985, a corporation in California applied for EPA approval to test such bacteria that would protect plants from frost. In 1987, under close supervision, company scientists in space suits sprayed a strawberry patch with bacteria, as nearly one hundred reporters looked on.

Mycogen Corp. has developed several genetically engineered agricultural pesticides -- biopesticides. Some plants produce substances in their tissues that are toxic to certain insect pests yet harmless to other insects, animals, and humans. One naturally occurring insecticide is a protein synthesized by the common soil bacterium *Bacillus Thuringiensis*. The protein is toxic to caterpillar-type insects such as cabbage loopers and tobacco hornworms. The protein causes jagged crystals to form in the leaves. When a caterpillar bites into a leaf, it gets a surprise mouthful of those crystals. The crystals rip up the caterpillar's innards. Some forms of the protein are also toxic to beetles and flies. Mycogen scientists cloned the gene in the bacterium responsible for producing the toxin. They inserted it into the cell of another bacterium that can be mass-produced. The cells are thereupon killed and encased in protective capsules that preserve the toxins until they are ingested by the target pest. The cells cannot spread or multiply -- because they are dead.

In another approach, the gene that manufactures *Bacillus Thuringiensis* is introduced directly into plants to yield strains of tomato, tobacco, and cotton plants that resist caterpillar pests. Also, plants can be made resistant to viral infections if they carry a gene encoding the otherwise harmless coat protein of the virus. In this fashion, it has been possible to engineer tomato, tobacco, and potato plants that resist a broad variety of plant viruses.

The roots of legumes (beans, peas, soybeans, peanuts, alfalfa, and clover) are naturally infected with the bacterium *Rhizobium* which has the ability to fix atmospheric nitrogen (i.e. convert nitrogen from the air into usable compounds in the soil). Due to their symbiotic coexistence with the bacteria, these plants are essentially "self-fertilizing." Efforts are being made to alter the *Rhizobium* bacteria so that they will infect other crop plants. These experiments also include the *Aspirillum* species which inhabit the root zones of tropical grasses and fix nitrogen used by these plants. *Aspirillum* is sometimes naturally present in the roots of wheat, corn, and sorghum. If its presence could be increased, more nitrogen would be fixed biologically by these plants.

The most common technique of gene manipulation in plants makes use of the plant disease agent *Agrobacterium tumefaciens*. In nature, susceptible plants that become infected by virulent strains of this bacterium develop tumors that are called crown galls. Crown galls actually result from the transfer and integration of genes from the bacterium into the DNA of the plant cells. The molecular biologist can profit from this circumstance by removing the genes responsible for causing disease from the bacterium and replacing them with DNA destined for a recipient plant. Through such infection, it has been possible to introduce new traits into a variety of species ranging from soybeans to apple trees. Current efforts to develop a blue rose (transferring into the rose a snip of DNA that is responsible for the color in blue petunias) are based on *Agrobacterium*- mediated gene transfer.

The EPA can take months and sometimes years to review health and environmental effects before it issues a permit for field trials. The concern of the EPA is that altered organisms released into the environment will cause health and ecological problems. Among the question-marks: Will altered bacteria released in the environment migrate to weeds or other plants and upset the ecological balance? Can bacterial toxins contaminate the food chain? There is also the danger that disease-causing organisms and pests, when faced with extinction, will mutate by themselves into new races capable of attacking previously resistant varieties.

Nobody really knows what would happen to our environment in a future world where hundreds or thousands of new strains of bacteria and plants created in the laboratory would be let loose. What if they started cross-breeding, generating new and unforeseen traits?

In a landmark decision, the Bush administration decided that genetically engineered food does not require the approval of Washington. The first genetically altered food to be sold to consumers is the "Flavr Savr" -- a new tomato developed by Calgene Inc. that can stay on the vine and ripen longer than ordinary varieties and stay fresh several days longer once it is on the grocery shelf. (Common tomatoes are picked while they are still green, stored for weeks in refrigerator trucks, and then gassed with ethylene to make them red.) But Calgene spent five years to get the approval of the Food and Drug Administration anyway, in order to allay the concerns of the general public.

Genetically Engineered Drugs

The methods of gene transfer via bacteria and viruses are being applied in the pharmaceutical industry to manufacture new drugs. One of the first success stories of genetically engineered drugs was the cloning of human insulin. Eli Lilly, Indianapolis, Ind. bursts open human cells and extracts the DNA which is then "cut" with special enzymes. The DNA pieces responsible for the manufacture of human insulin are put into a highly productive strain of the bacterium *Escherichia coli*. The bacteria are tested for the presence of the desired gene. It is sufficient to identify one single bacterium that has picked up the insulin gene. It is allowed to double. All of the cells in the resulting clone are identical and all follow the instructions in the human DNA to make insulin.

The final step is the industrial manufacture of billions of identical clones. It is done through a process called PCR -- polymerase chain reaction. The reaction occurs inside a machine called the DNA thermal cycler. PCR acts as a genetic zoom lens to magnify fragments of the genetic code. Each cycle doubles the number of clones. Starting out from a single copy of the strand of DNA, one next gets two copies, then 4, then 8, then 16 and so on. Each single cycle lasts only a few minutes; two hours of repeated cycling will increase the starting material by more than a million fold. The process is fueled by enzymes brought from a heat-loving microbe, *Thermus aquaticus*, that lives in natural hot springs and geysers. PCR can also be used for DNA fingerprinting (distinguishing one person from virtually any other be means of samples of their DNA brought from a single hair root or other material containing trace quantities of DNA), the study of human mummies, frozen mammoths, and unmineralized fossil plant tissue. The remains of Russia's last czar, Nicholas II -- executed in 1918 by the Bolsheviks -- were positively identified using a PCR test that matched DNA in the bones with DNA of members of the British royal family, who are distant relatives of the czar.

PCR was invented by Kary Mullis in 1983, a quirky genius working for Cetus Corporation, one of the first biotech firms. Cetus patented it and paid Mullis a $10,000 bonus. (Ten years later, Mullis could add to that sum a Nobel Prize and a check from the Swedish Academy of Sciences.) But Cetus was destined for trouble. In many ways, the fate of Cetus is symptomatic of the pitfalls and uncertainties in the fledgling biotechnology industry. Du Pont Co. took Cetus to court, trying to strip Cetus of its patent. The costly litigation took its toll on the small company.

Milestones in Biotechnology
1953 J. D. Watson and F.H.C. Crick discover the molecular structure of DNA deoxyribonucleic acid), the genetic key to heredity. It is a double helix.
1973 S.Cohen and H. Boyer make recombinant DNA, transplanting a toad's gene into bacteria, which then were able to reproduce the toad genes.
1976 Genentech Inc. formed by H. Boyer and R. Swanson.
1980 First biotech issue on Wall Street: Genentech floats a $36 million issue.
1982 Genetically engineered human insulin approved by the FDA. Developed by Eli Lilly.
1985 Genetically engineered human growth hormone (combats dwarfism in children) approved. Developed by Genentech.
1986 Genetically engineered alpha interferon (for treatment of hairy cell leukemia) approved by the FDA. The approval was later extended to other
1989 First human tests of gene therapy (to treat metastatic melanoma). The research team at the National Institutes of Health was headed by S.A. Rosenberg.
1993 First genetically engineered drug to treat cystic fibrosis. The drug helps liquefy congestion to clear airways and ease breathing. Developed by Genentech.

Can biotechnology products be patented? Under U.S. patent law, four classes of subject matter are allowed to be patented: a process, a machine, a composition of matter, or an article of manufacture. A natural product, such as a strain of *Rhizobia* that fixes nitrogen, is not patentable. The chemical that gives strawberries their characteristic smell and taste cannot be patented. But vitamin B12 which does not exist in nature in its pure form was granted a patent. A new plant hybrid can be patented. New species of plants, and seeds and tissue from such plants, developed by genetic engineering, can be patented. Even new animals can be patented -- in 1988 Harvard University was awarded a patent for a strain of genetically engineered mice, making them unusually susceptible to cancer, and thus well suited for cancer research.

The California company Systemix Inc. was granted a patent on purified bone marrow stem cells, often called the grandfather of all human blood cells. In a controversial effort, the National Institutes of Health (NIH) attempted to patent thousands of fragments of unidentified genes, but was rebuffed by the Patent Office.

Patents matter little if the courts will not uphold patent rights. Prior to 1982, fewer than one-in-five patents litigated were ultimately held valid and infringed. But in 1982, the Court of Appeals for the Federal Circuit was formed, with a mandate to handle appeals in all cases that raise a significant patent issue. With this new court, three out of four patents litigated today are ultimately found valid and infringed.

So, in the end, a federal court upheld the Cetus patents on the polymerase chain reaction. The David-and-Goliath struggle ended in victory for the small biotechnology company. But it was too late. The company was already on the ropes. Another calamity had struck.

Besides the gene amplification technology, Cetus' major product was Interleukin-2, a drug for advanced kidney cancer. Interleukin is a human protein. The interleukin story is a classic case of the difficulty of transforming the lab work of synthesizing proteins into practical products. Cetus had spent more than $120 million to develop it. Interleukin-2 had been given to more than 10,000 patients and was approved for sale in seven European nations. But it helps only a fraction of all kidney-cancer patients. In the summer of 1990, the Food and Drug Administration rebuffed the drug, objecting to the methods used by Cetus in its clinical studies.

Cetus stock collapsed. Three fourths of the capitalization of the stock of the firm was wiped out. The president and chief executive officer resigned. More than 10 % of the company workforce was laid off. One year later, the ailing company was swallowed by Chiron Corp. The PCR technology was sold to Hoffmann-La Roche, a Swiss pharmaceutical company.

The hopes and failures of Cetus illustrate the huge development costs and risks in the biotechnology field. Who pays for those costs? In the case of ultimate failure, the investors do. In the case of success, the consumers pay. The price of the product as it finally enters the market needs to be set high enough to enable the manufacturer to recoup the development costs and to provide additional return commensurate with the attendant risks taken.

In another use of Interleukin-2, a team of scientists at the National Cancer Institute got approval from NIH in January, 1989, to conduct the first test of gene therapy on humans. The test involved patients with advanced melanoma, a fatal skin cancer. The first patient at the Warren Grant Magnuson Clinical Center in Bethesda, Maryland, was infused with about half a liter of milky white solution containing almost 100 billion cancer-fighting cells bearing a marker gene. The cancer-fighting cells are white blood cells called tumor-infiltrating lymphocytes, or TILS. The cells have been soaked in Interleukin-2 to produce massive quantities of TILS.

The Biotech Boom, and the Food and Drug Administration

The risks in the biotech industry are huge. A few companies, like Genentech, Amgen, and Biogen have succeeded. But there is also a graveyard of biotech failures. How is it that the biotech industry is nevertheless able to attract large amounts of capital fueling a boom in biotech stocks? Many of the darling biotech startups on Wall Street are just a pie in the sky -- lofty promises but no products, no production, and no sales revenues. Financial analysts look at indicators like the "price-earnings ratio" of stocks. For most biotech stocks, no price-earnings ratio can be calculated, for the simple reason that there are no earnings. And yet, the investors keep flocking around the latest offerings.

It would be truly gratifying if somehow private capitalism could be harnessed to fund advanced research without relying on the government. But why does not the same formula apply to other industries? Why are there no initial public offerings on Wall Street in space research, in theoretical physics, or in psychology research, for that matter?

The answer is that the U.S. government, in the guise of the Food and Drug Administration, indirectly holds a protective hand over the biotech market, effectively guaranteeing multi-billion revenues for those drugs that pass muster . FDA is no doubt able to draw on the highest medical expertise in the land. But economically speaking, FDA has turned into a government machine for generating, and preserving, private drug monopolies. Those monopoly profits are driving the euphoria in biotech stocks.

To see how this works in practice, consider the case of Genentech's flagship heart drug TPA. TPA holds a commanding 70 percent share of the market for clot-dissolving drugs. It sells for $2,200 per dose. One of the competitors is streptokinase, manufactured by Astra AB and Kabi Farmacia AB of Sweden, selling for about $200 per dose. An extensive trial showed that TPA saves one in 100 more lives than streptokinase. In early 1995, Genentech won Food and Drug Administration approval to claim that TPA is superior, and to include key findings of the trial on the drug's labeling.

The result? Physicians will now feel it necessary to prescribe TPA rather than the cheaper alternative -- or they will expose themselves to the dangers of malpractice suits. The patients pay. Insurance companies pay. Medicare pays. There is no price competition.

It is a fallacy to believe that society can afford to eliminate the competitive mechanisms and to prescribe miracle life-saving drugs to everybody that needs them without regard to cost. The ballooning health costs of the country must be contained. The way to do it is not to regulate drug prices but to let competition drive prices down. In the particular case mentioned right now, the solution would be for the Food and Drug Administration to certify both TPA and streptokinase, but to refrain from putting its official stamp of approval on the precise efficacy of each drug.

Ethical Questions: Are We Playing God?

Almost every step of the development of new drugs based on gene splicing methods is dogged by controversy. About half a million American die from cancer every year. There is a desperate need for treatment. Many people think that the FDA uses too stringent criteria for the approval of new drugs.

Interleukin-2 is clearly effective in a subset of patients. Yet, Interleukin-2 is still labelled an "experimental drug" in the U.S.

Others hold that gene therapy is dangerous and should be stopped. Jeremy Rifkin and his Foundation on Economic Trends, an organization that monitors biotechnology issues, had sued to block the gene transfer that I mentioned a moment ago. Asked what, specifically, he is opposed to, Rifkin answers: "All releases of genetically engineered organisms into the environment. There's no risk-assessment science, there's no predictive ecology, no way for insurance companies to assess risk. It shouldn't be done because it's ecological roulette."

Obviously the nascent science of gene therapy raises questions beyond the more narrow issue of healing the sick. The day is coming awfully close when gene therapy will be available as a practical tool to enhance or "improve" the genetic makeup of man. One well-known novelist has already peered into this "brave new world" of the future: Aldous Huxley. What he saw was a nightmare.

As *Brave New World* opens, the reader is taken to the London Hatchery and Conditioning center of a future civilization, where humans are cloned and conditioned to fill their roles in a strictly ordered totalitarian state. There are five castes of humans, each with its distinctive genetic traits: alphas, betas, gammas, deltas and epsilons. Each caste consists of millions of identical twins -- "the principle of mass production at last applied to biology."

Writing in 1931, Huxley cleverly extrapolated from the biological knowledge available in his day. Human egg cells are fertilized in vitro, and are prodded by various chemical agents to "bud" in the incubators; through further "neo-Pavlovian" conditioning the human embryos are endowed with the desired traits of their caste. The procedures in the hatchery include electric shocks and "sleep-teaching." The epsilons are for menial labor; depriving the embryos of oxygen rids them of superfluous intelligence. (But Huxley never suggests that the hereditary structure itself would be altered -- the feats of modern genetic engineering no doubt would have sounded too far-fetched even to this bold thinker.)

Huxley's brave new world is set 600 years into the future. Technology is entirely enthroned, providing mankind with all conceivable bodily comfort. The hero, Bernard Marx, is an alpha-plus, destined for a leadership role in society. His girl-friend, Lenina Crowne, lives a complacent life without knowledge of want or pain. Driven by an impulse, they motor off in a rocket to a South American Indian reservation, where they meet a "savage" whom they eventually bring back to London. The savage cannot adapt to the artificial life of the brave new world, and ultimately hangs himself.

Huxley's novel deals with the revolution in the souls and flesh of human beings that would accompany the manipulation of the human race. He depicts a society that is robbed of freedom, beauty, and creativity. Lenina lives the life of a zombie, entertaining herself with "electronic golf," sex, and regularly lapsing into oblivion induced by a drug called "soma."

Is the U.S. today, in the early 1990s, at the threshold of the brave new world? Most people recoil at the thought of genetic engineering of humans. But we are willing to embrace "gene therapy" as a means of combatting genetic disorders. The evidence is accumulating that genetic degradation of some kind is associated with a range of diseases including cancer and Alzheimer's disease. If we shall be able to restore degraded genes, the step toward upgrading other genetic traits such as intelligence, memory, and other "alfa-plus" characteristics cannot be long. In brief, we shall be able to create a new super race of humans. Or, some sinister Frankenstein shall be able to create a race of subservient slaves.

Perhaps the writer of these lines is too old. I remember quite well the keen debate in the 1930s on the subject of "race hygiene" or eugenics. Hitler's scientists were dreaming of creating a pure Aryan German, rid of the bad genetic material of Jews and gypsies.

In the U.S., guidelines and regulations governing laboratory research involving genetic manipulation are being formulated by the Recombinant DNA Advisory Committee, which formally advises the director of NIH. The committee includes not only scientists but lay representatives and ethicists. When gene therapy on humans was approved for the first time, to treat metastatic melanoma, the committee stipulated that the treatment be limited to patients with life expectancies not exceeding ninety days.

The problem is not the treatment of terminal cancer. The problem is that a host of human afflictions and diseases, including arteriosclerosis and Alzheimer's disease seem to involve some kind of genetic degradation. If a pill each morning can delay the onset of memory losses in older citizens and beef up their mental abilities, what is there to prevent doctors prescribing the same pill for school children with "attention disorder syndrome"? Or for cocaine babies? Or for normal healthy babies? Where do we draw the line along this slippery slope?

Bibliographic Notes

Astounding news have been coming for some time from the botany and biology departments of our universities, and from fledgling start-up biotech companies. The day does no longer seem far off when man shall be able to "create" new crops, cattle, and seafood at will. These strides in biotechnology seem to herald a new age where most economic scarcities in the form of limited natural resources ("land" in the parlance of economists) would be removed. Instead, the limits will be set by the state of our current knowledge (all the time unfolding) and by our ability to convert fundamental biological knowledge to more efficient agriculture and aquaculture, and more efficient drugs.

The story of modern biotechnology is to a very large extent the story of Genentech, formed by Herbert Boyer, a researcher at the University of California at San Francisco. In its first years, the company had no laboratories nor a location of its own. Instead, the company signed contracts with researchers at the university to perform basic research. The biotech industry thus emerged as the result of the conscious strivings of a small group of academic people to commercialize university-based science and technology. In an interesting study, M. McKelvey *(Evolutionary Innovation: Early Industrial Uses of Genetic Engineering*, Department of Technology and Social Change, Linkoping, Sweden 1994) has traced the evolution of the industry from these early roots. Attempting to understand the conditions for evolutionary growth, McKelvey points at the presence of several competing research teams, and the synergy that developed between Genentech's own researchers and those of other biotech companies, both domestically and abroad.

References

For an introduction to the subject of health costs, see the articles under the heading "Controlling Health Care Costs" in Ref [11] and the references contained therein.

"The so-called Emmer wheat ..." Cited after J. Bronowski, *The Ascent of Man*, Little, Brown and Co., Boston 1973.

"The green revolution was spearheaded by..." N.E. Borlaug and C.R. Dorswell, "World Revolution in Agriculture," *1988 Britannica Book of the Year*, Encyclopedia Britannica, Chicago, 1988.

Box on Malcolm Brown: See also S. McCartney, "This 'Blob' Makes Tasty Spacesuits, A Nice Filet Mignon," *The Wall Street Journal*, January 6, 1994.

"In 1980 there existed about 35 000 different kinds of products ..." A.A. Boraiko, "The Pesticide Dilemma," *National Geographic*, Feb. 1980.

"Biotechnology is the use of living organisms..." See ref. [5], and also D. McCormick, "Biotechnology: Promise Redeemed," in ref. [8].

"The protein is toxic to caterpillar-type insects..." "Technology in the Year 2000," *Fortune*, July 1988.

"The most common technique of gene manipulation in plants..." J.L. Fridovich-Keil, "Genetic Engineering in the Garden," *1990 Britannica Book of the Year*, Encyclopaedia Britannica, Chicago 1990.

"One of the first success stories of genetically engineered drugs..." M. Momany, "Gene Busters," *Alcalde*, Jan./Feb. 1990.

"It is done through a process called PCR..." J.L. Fridovich-Kiel, "The Polymerase Chain Reaction," *1991 Britannica Book of the Year*, Encyclopaedia Britannica, Chicago 1991.

"Can biotechnology products be patented?..." See S.L. Misrock, T.E. Friebel and L.A. Coruzzi, "On the Patentability of Higher Life Forms," and also D.L. Parker, "Managing Patent Rights -- Concerns of the Biotechnology Venture and Investor," both in Ref. [13].

"Cetus stock collapsed..." M. Chase, "Chiron Agrees to Buy Cetus in Stock Swap," *The Wall Street Journal*, July 23, 1991.

"In another use of Interleukin-2, a team of scientists at the National Cancer Institute..." Steven A. Rosenberg and J.B. Barry, *The Transformed Cell: Unlocking The Mysteries of Cancer*, G.P. Putnam's Sons, New York 1992.

"Many people think that the FDA uses too stringent criteria..." R.K. Oldham, "In Cancer Research, Put the Patient First," *The Wall Street Journal*, Aug. 8, 1991.

"Jeremy Rifkin and his Foundation... Interview in *Bioworld*, April/May 1991.

"What he saw was a nightmare..." A. Huxley, *Brave New World*, Harper and Row, 1st ed. 1932, 2nd ed. 1946.

CHAPTER 9

Combustion, Fission, and Fusion

As the climax of the Gulf War approached, Saddam Hussein decided on a desperate and insane move: he had the contents of five Iraqi-flagged oil tankers dumped in the Mina al Ahmadi port in Kuwait. Also, the Iraqis opened the spigots of the oil-loading terminal off the Kuwaiti coast. A new kind of weapon was unleashed on the world: environmental terrorism. The ensuing Gulf oil spill was the largest spill in history, six to eight million barrels (the previous record: 4.2 million barrels dumped after the 1979 blowout of the Ixtoc well). It damaged a precariously fragile ecosystem. The Gulf is shallow and it is nearly enclosed. The average depth is a mere 110 feet. It takes 200 years to flush out and replace the stagnant gulf water.

Why were U.S. and allied troops battling Iraq in the first place? The answer is oil -- to fuel motorcars, to climatize homes, and as a source for the petrochemical industry. Modern technology cannot function without a readily forthcoming supply of energy. Indeed, the modern economy cannot function without energy. It feeds on kilowatt-hours.

War has been fought over oil before. In 1941, the U.S. had declared an embargo on oil to Japan. In October that year the Japanese government fell and the former War Minister Tojo assumed power. Tojo offered to withdraw from Indo-China in return for the U.S. resuming the oil shipments. Secretary of State Cordell Hull refused. Learning that his "final offer" had been rejected, Tojo set the date for the strike against Pearl Harbor one month later.

Rather than mounting an all-out attack on Moscow in the winter of 1941, Hitler opted for a major swing southeastward, through the Ukraine and the Donets Basin into the Caucuses. His goal: to open up German access to the Baku oil fields. Eventually he were to advance as far as Stalingrad. The German chemical industry produced synthetic gasoline to fuel the war machinery of the Wehrmacht, but the factories were hard pressed.

Are We Running out of Oil?

Is the U.S. running out of oil? The engineer will be tempted to answer in the affirmative, citing the dwindling amounts of known reserves. Even including the vast deposits slumbering under the permafrost in Alaska or beckoning the offshore drilling team in the Mexican Gulf, it is a statistical fact that the volume of known and recoverable reserves in the nation is dwindling. The United

Nation's *Energy Statistics Yearbook* figures that, at current rates of withdrawal, the proved reserves of the U.S. will be exhausted in 10 years. Production at Prudhoe Bay is already tailing off.

The seriousness of the situation is illustrated by the problems faced by Shell Oil Co. At one time, Shell was the No. 1 among U.S. gasoline retailers. It has been losing ground ever since. While other big U.S. companies are drilling and pumping extensively overseas where prospects are better, Shell, as Royal Dutch's U.S. subsidiary, is basically restricted by its parent to the U.S. Seeing a rapidly declining output, Shell is risking billions of dollars exploring expensive deep-water fields in the Gulf of Mexico, and on high-risk leases in Alaska's Chukchi Sea.

But to the economist, a country cannot really "run out" of anything. It is all a matter of price. At current oil prices, the oil industry in Texas is in the doldrums. But there is still oil deep down, of course. The largest U.S. oil field, located at Prudhoe Bay on Alaska's North Slope, has been supplying about one-fifth of U.S. domestic liquids production. Its production is channeled through the 800-mile Trans Alaska Pipeline System which terminates at the Port of Valdez, where tankers bring the oil to the lower 48 states. After some 15 years, production from Prudhoe Bay has peaked and has begun its inevitable decline. There is certainly oil in the nearby Beaufort Sea and Chukchi Sea. But each well there costs five times the expense of a deep-water hole in the Gulf. The sea is covered by a thick layer of ice. The prudent strategy to an oil company is to hedge against an uncertain future: to acquire some leases, to do some test drilling, and then just wait. Right now, a gallon of oil under the Chukchi Sea is worth more in the ground than at the wellhead.

Why do we have to depend on Kuwait and Saudi Arabia to supply us with oil, when our domestic reserves are certainly not yet exhausted? The answer lies in the interplay between technology and the endowment of natural resources. New technology lowers costs. Depletion of nonrenewable resources increases costs. It is an economic tug-of-war. For the greater part of this century, technology was the stronger of the two forces. A series of dramatic innovations caused the cost of prospecting for, drilling, and pumping a gallon of crude to come down. Among early innovations, one may mention the rotary system of drilling, magnetic exploration, and pipelines. More recent innovations include seismic methods of exploration, horizontal drilling, in-field drilling, enhanced recovery techniques, and offshore and deep-sea drilling. In the parlance of the economist, each such step shifted the marginal cost curve downward.

The tough adversary in this tug-of-war is depletion -- the fact that the most accessible deposits with the lowest costs are drilled first. As the available deposits are gradually being depleted, less accessible oil must be brought to the surface involving greater exploration, drilling, and pumping costs. For a while, innovations may lower costs. But eventually and inexorably, the costs of operating any given oil well will become prohibitive.

Most land-based oil pumping in this country is from old wells -- wells that were drilled many years ago. They are approaching the end of their economic cycle. By contrast, the vast fields in the Middle East, in the North Sea, and in Russia, are young. They have not yet reached the point of rapidly escalating unit costs. Even at the current depressed levels of oil prices, those countries can pump at a profit.

But many U.S. well owners cannot. The result to the nation is a bulging import bill. Today, half of all crude petroleum used here is bought from abroad. We are hooked on oil delivered from other nations. Our former self-sufficiency has been eroded. The much-heralded national Strategic Oil Reserve is a stop-gap measure only. During Operation Desert Shield there was considerable nervousness in international oil markets. Prices rose precipitously. To shield the domestic market, President Bush ordered oil to be released from the Strategic Reserve. But the quantities were too small to have any noticeable effect upon the markets. The truth of the matter is that we are at the mercy of the oil merchants of the Middle East.

The Hidden Costs of Oil

There is a price to be paid for being vulnerable. To protect our access to oil from the Middle East, the U.S. needs to ensure that Saudi Arabia and the small kingdoms along the Persian Gulf remain stalwart friends. Any subversion of the ruling political and geopolitical structure must be resisted. Hence the Carter doctrine and U.S. naval presence in the area. And hence, the Gulf War. Democracy does not matter. Stability matters.

In brief, there is an invisible security premium to be paid at the pump each time I fill up my car. One portion of the cost of the gasoline I pay in cash. Another portion I pay as federal taxes, to support our military commitments to keep open the flow of imports of oil into the country. How big is the invisible premium? The reader may get an idea of what is involved by dividing $50 billion (the reported cost of the Gulf war) by 2 billion barrels (U.S. imports of crude petroleum in 1990). That comes out as 25 dollars a barrel. In 1990, the invisible premium was about as large as the market price itself! Other years, the premium has been much smaller of course.

Is there no other way? Must we be so vulnerable? What can be done to bring down those stupendous oil imports? Well, there is the alternative of an import duty. The duty would drive a wedge between the world market price and the domestic price. It would raise the domestic price above the world market price. For a gallon of imported oil, the U.S. Treasury would collect the difference. For a gallon of domestic oil, the U.S. producer would collect the difference. The oil operations in Texas and Oklahoma and off the shores of California and Florida would benefit. At those higher domestic prices, a window of profit would be created between price and cost. Oil deposits that could not profitably be tapped at the lower prices would become attractive again. Domestic production would increase.

Both the Reagan and the Bush administrations rejected the idea of an import duty or fee on imported oil. In any case, isn't that socialism having the government fiddling around with prices?

The difficulty -- as I have discussed -- is that the free market price of imported oil does not reflect its true costs to the American consumer. A duty would not set the market mechanism aside. It would repair it, internalizing the invisible security premium, making the consumers pay the full cost up front rather than on their tax bills.

Economists have learned many lessons during the second half of this century. One of them is that nobody lives in an economic vacuum in the modern world. Acts of producers and consumers impact on other producers and consumers, either directly or via the environment. Production and consumption can carry important costs that are not reflected in the market prices. The solution to this dilemma is to repair the market mechanism. The repair work may be of different kinds, such as direct user fees imposed on consumers of products that inflict costs on society. Imported crude oil is such a product.

In the spring of 1989 the Governor of Texas appointed a task-force to look into the problem of oil prices. Our mandate was intriguing. One might have thought that the Governor wanted us to find out what could be done to raise oil prices. He didn't. There is something even worse to Texas drillers than depressed oil prices: unexpected price swings. Prospecting and drilling for oil is a long term proposition. It is a case of long term investment. If prices are volatile, the present value of the future income stream that the investment will generate is in doubt. Drillers need stable prices even more urgently than high prices. The problem that the Governor put to us was this: What can be done to stabilize the price of oil?

There exists a surprising answer to that question: let the importers bid on import quotas. Let the U.S. Treasury auction off the rights to import crude oil from abroad. Lower the volume of import quotas put up for sale if the world market price falls; raise it if the world market price increases.

A similar proposal was actually aired by the Clinton administration for textiles. Import quotas are a variation on the theme of an import duty. The sale of quotas would bring in revenue to the Treasury. They would raise domestic prices above the world market price. But the premium that domestic consumers would have to pay would not be a fixed duty. Instead, the premium would be determined in a free market and it would depend upon the volume of import rights offered up for sale by the Treasury in any given week or month. By making the Treasury a player in the market, effectively controlling the volume of imports of crude oil into the country, the Treasury would also, indirectly, be able to influence domestic prices.

For all practical purposes, the proposed scheme would work as a variable rate of duty rather than a fixed one. Only Congress can impose a duty on oil. The proposed arrangement circumvents the need to go to Congress every time one would want to change the rate and would instead delegate to the Treasury the daily management of oil imports into the country.

There are other costs of pumping and distributing oil that the consumer does not see. They include the costs of cleaning up after Saddam's eco-terror: controlling thousands of burning wells in the Kuwaiti fields, repairing burst pipelines, stranded tankers, and destroyed refineries. Much of the damage done to the environment cannot be repaired. It is irreversible. Oil spilled into water evaporates; it causes toxic fumes. The remaining mousse forms tar balls that sink and destroy the ocean floor including entire coral reefs -- the source of an abundance of marine life. Rare species of flamingoes and sea turtles find their nesting grounds infected. Remember that this is a part of the world that used to be known for the pristine beauty of its nature. Once, in the nineteenth century, Kuwait was a center for pearl-diving. Tourists were taken out to sea to watch the divers hurl themselves through the crystal clear waters, bringing up to the surface a rich harvest from the bottom.

The direct costs in dollars and cents of the *Exxon Valdez* oil spill in Prince William Sound in March of 1989 were steep enough. (Exxon paid about $2.5 billion on cleanup efforts and $1 billion to settle state and federal criminal charges.) But then there was the impact on millions of migratory waterfowl. Birds that dive for fish and eagles that eat oily carcasses were exposed. The spill in the sound threatened the biggest concentration of sea otters in North America.

The federal Clean Water Act holds companies liable for damages "for injury to, destruction of, or loss of natural resources". New legislation requires all new tankers to be equipped with double hulls. In the wake of the disaster, California passed the country's toughest and most comprehensive offshore oil-spill prevention plan. The law created a $100 million oil-spill cleanup fund, financed by a 25-cent tax on each barrel of oil.

But how do you pay for irrevocably destroyed flora and fauna, for damage done to planet Earth itself? There are obvious limits to what the price mechanism can accomplish. There is only one solution: sensible legislation, seeing to it that the greatest dangers are reduced. Fragile ecosystems in Arctic regions must be declared out-of-bounds. So, I think, should the outer continental shelf. Society must lay down tough ground rules of safety for drilling operations from rigs and platforms, and for the transportation and storage of oil in pipelines and tankers.

In the long run, implementation of such standards would increase world market prices of oil. That would be all for the good. Higher prices would make it profitable to look for alternative sources of energy, sources other than oil.

Coal and Natural Gas

Oil prices move in long cycles. The current world-wide glut carries the seeds of its own eventual reversal. Depressed prices inflate demand and hold back new drilling operations. A momentum of shortage will gradually arise, and the price of oil will eventually rebound. Conversely, elevated prices causes consumers to look for alternative sources of energy. Exploration and drilling will be stepped up. As larger volumes of oil hit the marketplace, prices fall again.

Electric utilities play a pivotal role in this dynamic process. Electric utilities account for more than a fourth of the use of fossil fuels in the U.S. Electric utilities burn oil (the lowest priced petroleum product, so called residual fuel oil), coal, or natural gas. Most utilities have facilities to burn at least two of these three fuels; many of them all three. Since utilities typically have excess generating capacity, they find it profitable to maximize the use of the cheapest fuel. When the price of a particular fuel is low, they will (after some time lag) step up the demand for that fuel.

More precisely, the managers of electric utilities adjust their use of alternative fuels based upon their expectations of future fuel prices. The world trembled as the OPEC quadrupled the price of oil in 1972-73. As oil prices rose further, reaching nearly $32 a barrel in 1981, oil became more expensive than coal, and the use of oil as a fuel by electric utilities dropped sharply. -- The sharp decline in oil prices after 1986 reversed the development, causing the electric utilities to step up their use of oil again.

Many look upon natural gas as a more attractive fuel than oil. Natural gas and oil are usually found together in the same geological strata, both presumably the end products of rotting and decomposing prehistoric giant ferns. But Professor Thomas Gold of Cornell University thinks otherwise. According to him, natural gas may not be of organic origin at all. He thinks that there are vast deposits of methane trapped deep inside the Earth, remnants from its creation. If he is right, the supply of natural gas is for all practical purposes infinite. We just need to drill deep enough.

In order to test Gold's theory, an exploration well was drilled from the lake of Siljan, Sweden, in 1987. Geologists know that the circular lake was formed thousands of years ago as an asteroid hit the Earth. Presumably, the outer crust of the Earth is thinner beneath the impact crater, and the inner core of natural gas -- if it does exist -- easier to reach. The experiment was discontinued one year and 21,780 feet later as the backers of the project ran out of money. The results were ambiguous.

Natural gas has been more victimized by government regulation than other fuels. Wellhead price regulation beginning in 1954 resulted in moratoriums on new service and shortages in interstate markets in the 1970s. The Fuel Use Act of 1978 restricted oil and gas burning in power plants and industrial boilers until the law was repealed ten years later.

Many observers think that natural gas will be the preferred fuel of the future. It is available in great abundance, and it burns cleaner than other fossil fuels.

Even now, natural gas is burned off as waste in much of the Mideast. The United States has centuries' worth of natural gas. There are thousands of independent producers, most of them located in the Southwest. The companies compete fiercely, which is one reason why natural gas remains so cheap. But the distribution system is a bottleneck. With today's glutted supplies, there is sometimes too much gas trying to force its way through too few pipelines. A dozen pipeline companies control the flow from wellheads to markets. The gas flows through the pipelines at a speed of 20 to 30 miles per hour. The transmission pipelines have compression stations every 50 to 60 miles to keep the gas flowing. The pipelines fan out from the Southwest to cover all of the nation, reaching the large distribution centers in California and on the East Coast. At the destination, the gas is received by large industrial users, an electric utility, or a local gas distribution company.

When burned, gas emits only about half the carbon dioxide of oil and coal. It emits less carbon monoxide and less sulfur. But the industry has not been successful in marketing its product. Electricity leads gas in both kitchen ranges and home heating furnaces. Gas just hasn't got the right user-friendly image. It is difficult to understand why. Gourmet cooks prefer gas. It heats faster. Whether you want your *Chateaubriand* steak rare or medium rare, the temperature can be more precisely controlled. Compressed natural gas (CNG) can be used to fuel cars. There are tens of thousands of fleet vehicles burning CNG on the road today. Amoco sells it from some of its service stations. But it has still to catch on.

It is difficult to compare the pros and cons of the three fossil fuels: oil, natural gas, and coal. Each fuel involves long networks of exploration, extraction, distribution and marketing. Each fuel requires its own combustion technology, delivering a unique mix of outputs: kilowatt-hours of course, but also carbon dioxide, carbon monoxide, sulfur, and a plethora of other byproducts with long complicated chemical names. Each of these byproducts has a range of "attributes", and they are all unwelcome and harmful -- negative attributes.

To make things even more difficult, these negative externalities and negative attributes are still poorly understood. Concepts like the greenhouse effect and acid rain only recently entered our vocabulary. The scientific understanding of these phenomena is still incomplete and evolving.

Don't Drink the Water and Don't Breathe the Air

Here are a few facts. The two main components of air are nitrogen (78.0 % by volume) and oxygen (20.9 %). Then there is the noble gas argon (0.9 %). Other constituent parts, like carbon dioxide, methane, and ozone, are present in small quantities only, none making up more than 0.1 %. These are the so-called greenhouse gases.

Perhaps carbon dioxide (CO_2) is the one component that has yielded the greatest surprises. Carbon dioxide is what makes the bubbles in carbonated soft drinks. It is what is inside the bubbles of champagne. It is colorless, odorless, and perfectly harmless to the human body. Yet, excessive levels of CO_2 may be the most dangerous of the greenhouse gases.

Carbon dioxide is part of a fine-tuned global ecological balance. Humans breathe in oxygen and breathe out carbon dioxide. Combustion engines burn hydrocarbons and release CO_2. All green plants and microorganisms do the opposite: through photosynthesis they convert carbon dioxide back into hydrocarbons and oxygen. This global balance has been disrupted for some time, and is getting gradually more out of whack. The burning of fossil fuels and of tropical rain forests spews more CO_2 into the air than the dwindling reservoir of green plants is able to break down again. The result is a pyramiding of carbon dioxide levels in the atmosphere. This is the greenhouse effect: increasing levels of greenhouse gases trap more heat from the sun inside the atmosphere. The average temperature rises -- global warming. The climate changes.

Since the creation of Earth, the global temperature has risen and fallen in long cycles. There have been ice ages. There was the age of the great dinosaurs when vast fern forests covered the land mass of North America. Some researchers believe that those ebbs and tides of the climate were accompanied by variations in the CO_2 levels. The composition of the atmosphere in times past can be established by examining air bubbles trapped in ancient layers of snow drilled deep from the Arctic ice. According to the "Gaia" theory (Gaia was the earth goddess of the Greek), the Earth and life on it are closely coupled. The composition of the atmosphere evolves in long cycles. An increase in CO_2 eventually leads to accelerated formation of plant life and plankton and thus set the stage for its own reversal.

The arrival of the Industrial Revolution changed the rules of the game. It is believed that the amount of CO_2 in the atmosphere has doubled since the early 1800s. Another doubling is expected to occur by the middle of the next century if no measures are taken to delay it. Most scientists agree that it will cause a rise in the average temperatures of between 2.5 and 8 degrees Fahrenheit. This may not seem like much, but would mask pronounced changes in many areas. The warming will be most intense in high latitudes, the world's grain belts will become drier, and rises in sea level will affect people living in deltas and low-lying coastal areas.

Acid rain develops as the burning of fossil fuels pours millions of tons of sulfur and nitrogen into the air. Most of these emissions are from power stations and the steel industry. In contact with air, sulfur and nitrogen are oxidized to form sulfuric and nitric acids. An acidic haze is formed. When it rains, the droplets pass through the haze and acid rain falls to the earth. Winds carry acid clouds over great distances. Over half of Canada's sulfate precipitation is believed to have originated in the U.S. Acid rain kills forests and lakes and rivers.

The Clean Air Act was passed in November 1990. It amends earlier legislation dating back to 1970. The new measures mark the first time the nation has taken action to stop acid rain. The Environment Protection Agency (EPA) is mandated to enforce the controls. Tailpipe exhaust from all motor vehicles has to be reduced. New cleaner gasoline is to be developed. The standards will be particularly tough for fleets of taxis or buses, encouraging the use of non-gasoline fuels such as methanol and natural gas. Coal-fired power plants must cut the amount of sulfur emitted, either by installing smokestack scrubbing devices or by switching to cleaner-burning low-sulfur coal.

The California Air Resources Board has adopted the nation's most ambitious emission control plan. By the end of the century, all new cars sold in the state will have to meet the most stringent exhaust standards; a rising fraction of all vehicles must emit *no* polluting fumes at all. The Big Three auto makers are all preparing to bring their own electric vehicles to market in time for the state's 1998 deadline.

The benefits of environmental regulation should be obvious to everybody. But I need to pause for a moment to discuss briefly the concurrent cost of such regulation. Assuming that the marketplace at all times correctly identifies and employs the cheapest mix of energy sources, the purpose of the regulation is to bring about a mix that the market would not have chosen of its own volition -- it would not have chosen it because it is more expensive. High tech catalytic converters to be fitted in the tailpipe of your car cost money. So do scrubbers to be installed in smokestacks. It is more expensive to buy -- and to mine -- low sulfur coal than to use the plentiful and dirty high sulfur variety. And so on. The regulation of any one technology therefore has an implied cost. In the language of economists, it has an "imputed cost" or a "shadow cost". It is measured by the net increase in total energy costs paid by the consumers.

The obvious alternative then presents itself of charging the shadow cost up front, as an environmental fee rather than an implied cost. Here is an example: leaded versus unleaded gasoline. Lead is a serious poison. To society, leaded gas has a shadow cost over and above its manufacturing cost. How do you make the individual consumer respect that cost? Consider these two alternatives: (1) In July 1984, the EPA proposed a 91 % reduction in the amount of lead permitted in gasoline. A few years later, a total ban followed. Gas stations were required to install new equipment that made it mechanically impossible to fill unleaded gas into the gas tanks of all new cars. (2) An environmental fee of a few cents is levied on each gallon of unleaded gas sold. After a transitional period, the fee is increased so that it exceeds the manufacturing cost differential. At that point, all sales of unleaded gas will automatically cease.

I am not necessarily proposing the second alternative. I mention it to explain how an environmental fee works. In this particular example, a straight prohibition does the job just as well. In many other cases, that kind of drastic action is not desirable. Nobody advocates that we cut the use of fossil fuels down to zero. What is needed is to slow down the consumption of oil, gas, and coal and to accelerate the development of alternative fuels (see box). To fine-tune the desired reduction, environmental fees are superior to direct regulation.

The problem with direct regulation is that it forces everybody to comply rather than allowing the individual energy consumer to seek out his optimal adjustment. An environmental fee still leaves the consumer with options such as installing various kinds pollution abatement equipment, or switching to other sources of energy. Or even doing nothing, just paying the fee as charged! As each consumer minimizes his individual costs, total costs to society will also be minimized.

Environmental fees can take the form of direct (excise) taxes. They can also take the form of tradable permits. Read on.

The Clean Air Act contains an innovative pollution-trading system in which electric utilities that make extra-deep pollution cuts get credits -- so-called emission allowances -- that they can keep or sell to other power plants. This arrangement sets up a direct cash incentive for the polluters to develop or acquire technology that reduces emissions. Those who fail to do so will have to buy allowances from other more successful utilities. The environmental externalities become "internalized" and show up directly in the profit and loss statement of the utility.

The allowances are traded at the Chicago Board of Trade. The contract unit of trading is defined as "25 one-ton sulfur dioxide emission allowances". Contracts in nitrogen oxide are also in the works. Trading includes both spot deals and futures contracts. In the futures market an electric utility is able to contract already today for an emission allowance to be delivered 3 months, 6 months, or 9 months hence. The futures market is also open to pollution arbitrageurs, that is to pure speculators (just as there are arbitrageurs in the markets for sugar futures or sugar futures). "Arbs" pump up the total volume of transactions and they cause futures prices to vary smoothly with the length of the contract. They help translate all available market information into current

A Checklist of Alternative Energy Sources (Alternative to Fossil Fuels)	
Hydroelectric	A clean and in many ways ideal source of energy. Well established technology. But few waterways suitable for additional development are left in the U.S.
Solar	Photo voltaic cells or film: One or several layers of semi-conducting material bathed in sunlight create voltage that drives electrons through a circuit.
Agricultural biomass	Ethanol is manufactured through the fermentation of corn, sugarcane, or agricultural waste. Used alone as a motor vehicle fuel or as a gasoline additive ("gasohol"). No toxic exhausts.
Geothermal	Geothermal heat-pumps: Water pumped down toward the earth's interior picks up heat for recovery at the surface.
Wind	Wind farms: Clusters of huge propellers on towers spin generators and produce electricity. Computer generated turbine blades, variable speed rotors.
Tidal	The Waveberg (invented by Canadian John Berg). Looking like a huge octopus, it has hinged arms attached to a central pod anchored to the bottom. As ocean swells pass, the arms rise and fall, driving a pump.

price quotations. The result should be dramatic: for the first time can the public now get a precise measure of the cost of environmental rules.

The new trading vehicle represents a triumph of economic theory. Economic science has not had too many such success stories, so we better take note. It all goes back to Ronald Coase, the University of Chicago law professor who got the 1992 Nobel prize in economics for his work on the importance of establishing clear property rights. He realized what nobody had understood before: that the competitive market mechanism can actually be harnessed to correct any environmental damage such as pollution once the ownership of the environmental resources has been defined and the resources are made tradable.

In 1995 the first plants began receiving pollution credits from the EPA, each of which allows one ton of sulfur-dioxide pollution a year. Companies will get credits for between a third and half of what they used to pollute. The EPA will also sell a few thousand extra credits in an annual auction. The price of the pollution permits will be set in the marketplace, by supply and demand.

The reader will remember that I have already proposed auctions on the rights to import oil to this country. Now I am proposing auctions on pollution rights. And I will go further. A wide range of environmental externalities that distort the free market economy can be corrected by government allocation of rights and the sale of those rights at public auctions. The cost charged the final consumer will thus includes an environmental fee.

Take, once more, the case of automobile exhausts. Imagine the following organizational framework: The EPA formulates a target for the annual release of hydrocarbons, nitrogen, and sulfur from motor vehicles in each state. It issues emission permits up to the determined limit. The permits are sold at monthly auctions. The buyers are dealers -- middlemen -- who resell the permits to individual inspection stations. At the time of my regular state inspection, an optical scanner analyzes the exhaust from the tailpipe of my car. I pay the pollution fee at the going market rate.

We need to develop energy technologies that work in harmony with the environment, rather than against it. One part of such a program is the control of existing technologies and the imposition of environmental fees, as explained. But the most important part is the development of new environment-friendly technologies (see box on previous page).

Perhaps such a new technology is fermenting corn into ethanol. Perhaps it is solar panels. Perhaps it is geothermal energy. I do not know. What I do know is that environmental fees change the relative prices of energy. Hydrocarbons becomes more expensive, relative to alternative fuels. Consumer demand for the alternatives is boosted. The profit prospects of firms in the alternative energy industry improve. Initial public offerings (IPOs) on Wall Street by such firms are more enthusiastically received. The splendor of the creative genius of American capitalism is marshalled.

Nuclear Fission and Nuclear Fusion

It remains to say a few words about fission and fusion. The fuel used in all nuclear reactors in the U.S. is enriched uranium. The energy comes from the breakup or fission of the uranium nuclei. In a light-water reactor, water is circulated through the reactor's fuel core to keep it from overheating. Elaborate safety systems have been designed to anticipate and prevent a loss-of-coolant accident. In March 1979, at Three Mile Island, a pump that provided feed water to the core of the reactor clogged while the reactor was operating at full power. Radioactive water flooded the plant. The cleanup cost almost §1 billion. But the reactor never spewed radiation clouds as did the Soviet Union's 1986 Chernobyl disaster. The reactor vessel and the 20-story concrete containment building remained intact.

Also, there is the problem of nuclear waste. What do you do with the spent fuel rods removed from the reactor when they have been used? One option is vitrification and storage in underground caverns. The nation still has to devise a master plan for the permanent storage of its nuclear waste.

Public anxiety about nuclear energy makes it extremely difficult to get a commercial operating license (issued by the Nuclear Regulatory Commission) for a new reactor in the U.S. today. Several finished plants remain idle, unable to operate because local authorities will not cooperate in the demonstration of workable emergency evacuation plans. All nuclear plants ordered since 1974 have been canceled. No new ones have been ordered. In one word: the nuclear program in the nation is stalled.

I should point out an alternative to the conventional uranium fuel: weapons-grade plutonium. France and Japan are pegging their hopes for cheap energy on it, building so-called fast-breeder reactors. And North Korea has been looking at it too, possibly with more sinister intentions. It is an invitation to international terrorism.

The economics of nuclear power is the economics of avoiding disasters. As we have seen, most forms of energy -- being a form of an economic "good"-- comes together with the production of economic "bads": pollution, the greenhouse effect, and so on. In the case of nuclear energy, the bads (negative externalities) are of a particular kind. They are risks. Actually, they are very, very small risks of very, very grave occurrences. The societal management of nuclear power is an instance of risk management.

To the reader of this book, the subject of risk should be familiar by now. The risks of new product development, of distribution, and of marketing. The need for management to accept risk as part of its normal working conditions. The need to contain risk. But this time, it is a different kind of risk: the risk of bads rather than goods. How can an integrated oil company, like Exxon, manage the risk of oil spills? Minor spills occur fairly frequently, large spills only rarely. Each spill has its attendant cost, to the company, and to society. The company needs to contain the actuarial value of its total annual risk exposure. So does society. The company will find it worth while to institute a policy of risk avoidance, including a suitable mix of safety measures, and contingencies of spills that do occur. So does society.

But nuclear risks pose problems of a different order. The calculations that I have just indicated break down. What is the actuarial value of the total risk exposure of a nuclear installation? There is a mathematical problem here: multiplying a very, very small risk with a very, very large cost. It is the problem of multiplying a number close to zero by another number close to plus infinity. Such an operation belongs to the mathematical discipline of infinitesimal calculus. The solution is indeterminate. There is no guarantee that the result is finite. The actuarial value of the risk (the "mathematical expectation") may not even exist!

Nor will there then exist any optimal policy of managing the risk. In simple words: the risk cannot be managed. It cannot be handled. There is just one way to deal with it: to avoid it. I do not consider nuclear energy a viable energy option.

Nuclear fusion is the process by which energy is created inside the sun. At its core, the sun traps hydrogen spiraling inward in a gravitational vise of crushing strength. The hydrogen nuclei collide and fuse. Mass is converted into energy, or, in the idiom of Albert Einstein: $E = mc^2$. The released energy radiates into space. To achieve fusion, scientists must re-create the extreme conditions that exist inside the sun. It is done inside a doughnut-shaped reactor called the Tokamak (a Russian acronym). Fusion researchers claim that they are close to "scientific break-even": the point at which the energy produced in the Tokamak would equal the energy employed to heat it. That would still leave the researchers with the daunting challenge of achieving a *commercial* break-even.

The economist who tries to penetrate these matters quickly runs into an unexpected difficulty. Fusion scientists live in a world of make-believe (or, they are singularly unscrupulous in motivating their funding requests). Since the 1950s, the fusion researchers have intoned the siren song of endless energy. The U.S. pours close to half a billion dollars into fusion research every year. Penthouse magazine publisher Guccione blew §16 million on a commercial venture to build a fusion reactor. Two now infamous chemists, Fleischman and Pons, sweet-talked the University of Utah and the state of Utah to put up millions to produce "cold fusion". But commercial application is as distant as ever.

Yin and Yang

So, where do we stand? The environmental costs of energy are staggering. New kinds of environmental damages are discovered all the time. There seems to be no end to the litany: damages to the air, to the water, to the soil, to the flora and the fauna, to our health, to the climate, even genetic damage to generations unborn. It is as if the Earth itself were cringing in pain, crying out ever more desperately.

Until quite recently in historic times man was part of a complex system of ecological balances. What he captured and destroyed, nature replenished and rebuilt. The industrial age disrupted this precarious symbiosis between man and his environment. Lately, the system has been careening out of control. Rising material living standards for everybody requires stupendous amounts of energy. Blind to the consequences, we are raping the Earth.

In many ways, modern civilization has alienated itself from nature. Nature is something that people watch on TV, from the comfortable vantage point of their reclining chair. It is scenery. Many seem to regard it as a kind of fungible raw material, useful as an input into manufacturing processes.

I think it is a matter of attitude. We need to recover the sense of humility that people used to feel in the past -- thankfulness for our daily bread and the bounty that nature and the environment provides. The old Chinese had a word for the complementary and harmonious interplay between pairs of opposites: *Yin* and *Yang*. Yin is the Earth, nourishing and yielding. Yang is man. All parts of the universe should be attuned in a rhythmical pulsation. Any energy policy worth its name must recover the harmony between Yin and Yang.

Is this possible? Is it possible to live a modern life and yet return to nature as much as we take from it? In order to find out, the Space Biospheres Ventures built a "biosphere" in the Arizona desert, 30 miles north of Tucson. The three-acre glass-and-steel complex includes a tropical rain forest, savanna, marsh, ocean, desert, farm, and human habitat. In September 1991 four men and four women were sealed inside, together with some 3,800 species of plants and animals. Their mission: to stay inside for a total of two years, isolated from the outer world, testing an artificial self-sustaining life support system. A typical breakfast: papaya crepes, goat-cheese yogurt, multigrain granola, home-baked bread, and assorted fruit. The living quarters are duplex apartments with computers, video-link telephones, TVs and VCRs.

A year later, the venture was riddled with controversy. A mechanical scrubber had been used to remove excess carbon dioxide from the biosphere. The air was replaced when the seals were found to have leaked. One of the Biosphereans had briefly left the enclosure to have surgery on a cut finger. The entire project had become a kind of amusement park, replete with a Visitor Center, snack bars, and gift shops. But behind the hoopla remained a deadly serious question: Can man live at peace with the environment?

Bibliographic Notes

Brutally awakened by the oil crises of the 1970s, economists have since spent much effort not just investigating the world market for oil, but also developing and consolidating the entire field of energy economics. It is now one of the better developed disciplines of economics.

Initial progress centered around the issue to what extent various sources of energy can be substituted for each other in the production process. Also, there is the question whether it is possible to reduce the overall use of energy by adopting energy-saving technologies and energy-saving machinery and other fixed capital. In order to deal with these matters, economists learned how to formulate mathematical production functions that use energy, capital, and labor as productive factors and to estimate such functions statistically.

With these building blocks at hand, energy economists turned to constructing equilibrium models of the markets for fossil fuels, both partial models for these markets alone, and imbedded inside the setting of an entire national economy. (For the U.S., A. Manne constructed the ETA-MACRO, a model of two-way linkage between the energy sector and the balance of the economy. For references, see below.) With such tools, economists became able to answer questions about how some given change in energy supplies would be propagated throughout an economy. On a global scale, one could determine how a cut in the world supply of oil (engineered by the OPEC in the form of production quotas), initially raising the price of oil, would eventually cause energy users to switch from oil to other sources of energy.

As international oil prices slowly dropped again during the latter half of the 1980s, economists became concerned that the temporary glut of oil only masked a future energy shortage. Many advocated slapping heavy excise taxes on oil and other fossil fuels, thus slowing down the gradual exhaustion of precious reserves. Such a tax is often called a "Pigou tax" after the British economist A.C. Pigou. Already in the beginning of the century Pigou had suggested that the government might enhance the "social welfare" in a nation by imposing excise taxes or user fees on goods and services that inflict indirect costs to society.

Incorporating any given array of Pigou taxes into the market prices to be determined in an equilibrium model, it is possible to solve for the effects on all demand and supply decisions in the economy. Conversely, starting out with some *a priori* targets for the supplies of fossil fuels and other sources of energy, one can solve for the taxes that will fulfill those targets. In particular, such calculations have been made to solve for the hypothetical "world carbon taxes" that would contain CO_2 emissions within set limits. (See A.S. Manne and R.G. Richels, *Buying Greenhouse Insurance - the Economic Costs of CO_2 Emission Limits*, MIT Press, Cambridge, 1992.)

References

"It damaged a precariously fragile ecosystem," Sharon Begley, "Saddam's Ecoterror," *Newsweek*, Feb.4, 1991.

"...the problems faced by Shell Oil Co." C. Solomon, "Shell, a Fallen Champ of Oil Industry," *The Wall Street Journal*, August 30, 1991.

"The largest U.S. oilfield, located at Prudhoe Bay..." L.L. Lamberton, "Developing Oil and Gas Worldwide," *The Lamp*, Fall 1992.

"...the Governor of Texas appointed a task-force..." *Stabilizing U.S. Oil Prices*, A Report Prepared for Governor William P. Clements, Jr., August 7, 1989, by a Task Force on Oil Price Stabilization, The University of Texas at Austin.

"There exists a surprising answer..." S.L. McDonald and S. Thore, "Auctioned Quotas to Limit U.S. Imports," *Journal of Petroleum Technology*, December 1989, pp. 1332-1334.

"Electric utilities play a pivotal role..." S. Thore and R.J. Gonzalez, "On the Determination of the Equilibrium Price of Oil," presented at the Tenth North American Conference of the International Association for Energy Economics, Houston, October 1988.

"According to the 'Gaia' theory..." See J. Lovelock, *The Ages of Gaia*, Bantam, 1990 and *Scientists on Gaia*, ed. by S. H. Schneider and P.J. Boston, MIT Press, 1991.

"This may not seem like much, but would mask..." K. von Moltke, "Global Environment - a Planet in Stress," *1991 Britannica Book of the Year*, Encyclopaedia Britannica, Chicago 1991.

Checklist on Alternative Energy Sources, tidal energy: J. Kurtz, "Closing in on sea power with a three-armed bandit," *The New York Times*, August 25, 1991.

"At the time of my regular state inspection, an optical scanner analyzes..." For calculations of the market values of rights to emit carbon dioxide under a hypothetical regime of international regulation, see A. Manne and R. Richels, "Global CO_2 Emission Reductions -- the Impacts of Rising Energy Costs," *The Energy Journal*, 1991.

"In March 1979, at Three Mile Island..." S. Shulman, "Legacy of Three Mile Island," *Nature*, March 1989.

"The U.S. pours close to half a billion dollars into fusion research..." R. Herman, *Fusion: The Search for Endless Energy*, Cambridge University Press, New York 1990.

"But behind the hoopla remained a deadly serious question..." E. Jarolim, "Yuppie Utopia: Touring the Biosphere," *The Wall Street Journal*, May 26, 1992.

CHAPTER 10

Ventures and Start-up Companies

Athens and Sparta were city states. Hamburg and Bremen were German city states and are to this day independent *Lande* in the Federal Republic of Germany. The Greek word "polis" (plural poleis) means city or city state. A technopolis is a modern form for city state based on high technology -- a city of high tech corporations. The first technopolis was Silicon Valley. Others are Route 128 (drawing on the expertise of the Massachusetts Institute of Technology), the Research Triangle in North Carolina (formed by Duke University, The University of North Carolina at Chapel Hill, and North Carolina State University at Raleigh), Sophia Antipolis at the French Riviera, Tsukuba City in Japan, and the Shenzhen Special Enterprise Zone in the People's Republic of China. A number of other technopoleis are now emerging in the United States and in other parts of the world.

There are obvious advantages to craftsmen and corporations to locate in close proximity to each other. The goldsmiths and silversmiths in Florence flourished on the Ponte Vecchio, the glass blowers on the island of Murano outside Venice. The tailors in London had their quarters on Savile Row, the bankers in the City. The English textile industry was concentrated around Manchester, the German steel industry in the Ruhr valley. So-called location theory tells us that if there is just one ice-cream parlor in town and you plan to open up a second one, you should open up next door to the one already in existence.

Silicon Valley: the First Technopolis

Silicon Valley, a thirty- by ten-mile strip between San Francisco and San Jose, got underway in 1957 when Robert Noyce and seven other brilliant young engineers quit Shockley Semiconductor Laboratory (William Shockley, co-inventor of the transistor) to launch Fairchild Semiconductor. These cofounders later split off to launch over eighty semiconductor firms in Silicon Valley over the next thirty-five years. Eventually, there were to exist over 3000 microelectronics firms in Silicon Valley; two thirds of them would have less than 10 employees.

Which were the key factors in the rise of Silicon Valley? First, there was an availability of technical expertise. In the formative years, the proximity to Stanford University was important. William Hewlett and David Packard were students in electrical engineering at Stanford when they started production in a garage behind their rooming house in Palo Alto. The University had started a research park -- the Stanford Industrial Park -- and Hewlett-Packard was among the first firms to locate in it. Second, there arose in the Valley at an early point the required infrastructure: suppliers, markets, and financiers. Third: job mobility. Fourth: personal networks facilitating the exchange of ideas and technological knowledge.

To illustrate how these factors worked in practice, consider the case of Apple Computer. Stephen Wozniak had built the first prototype in his garage, while he was still employed full-time by Hewlett-Packard. Wozniak and Steven Jobs had met at the "Home-brew Computer Club," composed of people in Silicon Valley who were avid computer hobbyists. Wozniak had built the computer around a Motorola 6800 microprocessor which Hewlett-Packard offered to its employees at a discount. Jobs had returned from a stint in India, had been into fasting and primal screaming, and worked on and off at Atari. Jobs convinced Wozniak that they could make printed circuit boards and sell them to computer hobbyists. Many of the members of the club only attended the meetings in order to snap up an opportunity to sell or buy electronics parts. One of them had opened a store in Mountain View called the Byte Shop, selling electronics components and computer kits. The owner of the store was impressed by Wozniak's product and ordered 50 of them. Wozniak and Jobs went to work in Jobs' sister's bedroom in April of 1976. The moment of creation. Four years later, Jobs was worth 165 million dollars.

Obviously, there are "multipliers" at work in the technopolis. As firms expand, they create jobs and wages and local purchasing power. As a manufacturer of computer hardware gets orders, it will itself place orders on semiconductors and other components required for assembly. In addition, there are powerful so-called positive externalities: the growth of one corporation builds infrastructure and stimulates the emergence and growth of other corporations. A technological environment develops. The environment attracts new talent, technicians, engineers, managers, and entrepreneurs. It is a snowballing mechanism.

The technopolis feeds on the development of new technology -- a continuous flow of new products and new designs that leave the laboratories and enter the marketplace. Some of these new products and new designs are commercialized by the developers of the products themselves. Others are transferred to spin-out companies or to competitors. The transfer of new technology from the laboratory to commercialization is the life blood of the technopolis. Commercialization occurs within existing companies or by new start-ups.

Can a technopolis be planned? Other states have looked upon California with envy. Planners and government bureaucrats from all over the world have travelled to Santa Clara and Sunnyvale and San Jose and Palo Alto to see what it would take to transplant the phenomenon of Silicon Valley to their home state. They all returned intrigued by the concept of the industrial park. Can local government foster a technological environment by building industrial parks or "incubators" that hold fledgling companies under their arms while growing up?

I shall return to these questions, but let me first outline a family of various institutional arrangements all promoting the transfer of technology and the formation of new companies. They include: R&D consortia, Federal laboratories, business incubators, venture capital firms, and the small business innovation research program passed by Congress.

The R&D Consortium

High tech research is expensive and it takes time. Sometimes it takes giant firms to be able to finance and man the vast R&D departments that are necessaryto stay abreast in the race to develop the technology of tomorrow. In 1980, Japan declared its intent on becoming the world leader in innovation and technology. What can medium size U.S. firms do to meet the global competition in the research laboratory?

One answer is the R&D consortium, which enables several companies to pool their R&D efforts. The consortium allows the member companies to share a scarce resource: qualified researchers. Duplication of individual research efforts is avoided. The result is a more efficient use of each member's R&D budget. For a long time, joint R&D ventures were considered illegal under the provisions of the antitrust laws contained in the Sherman Act (and the subsequent Clayton Act). But in 1984, Congress passed the National Cooperative Research Act which permits companies legally to engage in joint research activities. The activities are restricted to pre-competitive research. Product involvement, manufacturing or distribution is not permitted. One of the earliest consortia is the Microelectronics and Computer Technology Corporation (MCC) located in Austin, Texas.

In 1981 the Japanese had announced the creation of the so-called Fifth-Generation Project, which had as its expressed goal to make Japan the world leader in advanced computer technologies and "fifth generation" computers by the 1990s. MCC was born in the wake of this announcement, to respond to the Japanese challenge. It is a private-sector cooperative research venture, founded by three manufacturers of semiconductors (AMD, Motorola, National Semiconductor) and seven computer companies (Control Data, Digital

Equipment, Harris, Honeywell, NCR, RCA and Sperry). The semiconductor companies were smarting from Japan's recent success in capturing the RAM (random access memory) market. The computer manufacturers were pressed into action by their ongoing competition with IBM. MCC was to be active in the fields of software technology, semiconductor packaging, computer-aided design, and advanced computer architectures. Each company owns one share of stock, and agrees to share the cost of the research programs in which it participates. MCC produces "generic" technology, that is, innovations that could be incorporated into a wide range of products, such as new ways to interconnect semiconductor chips. But it cannot produce any products or become involved in manufacturing. The innovation that comes from MCC must be transferred back to the shareholders who can incorporate it into their own products and eventually carry those products competitively to the market.

MCC was the brainchild of W.C. Norris, founder and CEO of Control Data Corporation. For some time, the Reagan administration had been sending signals to industry that cooperative research programs would not necessarily be considered illegal. The proposed consortium got the go-ahead by the Department of Justice in December 1982, two years before the National Cooperative Research Act had been passed. Bobby Inman -- a retired four-star admiral and former deputy director of the CIA -- was selected to be MCC's first president and CEO.

In many ways, the formation of MCC can be seen as a major national experiment aimed at developing new technology, strengthening U.S. competitiveness, and beating back Japanese competition. At the time, there were few precedents in the Western world. Hundreds of other consortia have followed later, but MCC remains the most visible of these efforts. How did it work out? Success or failure?

Since its formation, MCC has undergone structural, cultural, and operational changes. By mid-1993 it was made up of 22 shareholders and scores of associates, small business associates, and university affiliates. It has transferred several hundred technologies back to its members, to universities, and to private corporations. But critics argue that MCC has produced minimal results and that the money spent by the members could more profitably have been invested in their companies' own R&D. Lately, several of the shareholders have left the consortium.

To discuss these matters, the concept of the "technology transfer gap" is helpful. The gap is between potential commercial application and actual successful commercialization. Physically, the gap may take the form of complete technical specifications, patents, drawings, and blueprints that have not yet found any commercial takers. More fundamentally, the gap is in peoples' minds: the gap between what the technologists have seen and accomplished, and what the businessmen have yet to find attractive and potentially profitable. The difficulty facing a research consortium like MCC is that it is expected to generate new high-quality new technology *and* it is expected to translate the new technology into successful commercialization. In other words: transferring the new technology back to the MCC members.

Unfortunately, one knows very little about these matters. Technology transfer is a very, very difficult thing to accomplish in practice. One of the lessons from MCC is that even the members themselves of a research consortium may take a sceptical and detached attitude toward technology developed in their own laboratory. If MCC does not actively "push" the new technology back to the members, and if the members do not actively "pull" it in, the transfer will never occur. In a book titled *R&D Collaboration on Trial*, D.V. Gibson and E.M Rogers provide a long list of transfer strategies used by MCC, including such "push" arrangements as the assignment of in-house member representatives, member visits, "company days," membership workshops, and the "PartnerMart." On the "pull" side, MCC is presently encouraging third-party licensing, vendor-company participation, spin-out technologies and start-up companies. A for-profit subsidiary named MCC Ventures supports the licensing and commercialization of new technologies whether developed at the consortium or elsewhere.

Other private-sector research consortia. After the passage of the National Cooperative Research Act, the number of consortia registered with the U.S. Department of Justice grew rapidly. They now include several hundred R&D consortia involving many thousand U.S. companies. Examples: The Semiconductor Research Corporation, the Center for Advanced Television Studies, The Plastics Recycling Foundation, the Consortium for Superconducting Electronics.

SEMATECH (SEmiconductor MAnufacturing TECHnology initiative) is a not-for-profit industry-government consortium, developing the tools and equipment necessary to manufacture semiconductors. Sematech is the Pentagon's flagship project in industry government cooperation. It was established to bring U.S. semiconductor manufacturing methods into parity with the Japanese. The government funding is channeled through The Defense Advanced Research Projects Agency (ARPA), a little known arm of the Department of Defense that funds high technology research.

A bungled consortium. In June of 1989, IBM and six other U.S. based electronic system and semiconductor producers formed U.S. Memories, a consortium designed to help the U.S. regain its competitiveness in the production of dynamic random access memories (DRAMs), which are critical components of most computer systems. The notion was that only a big, vertically integrated U.S. company could hope to compete with Japanese conglomerates. But some of the most important companies -- Apple Computer and Sun Microsystems -- balked at the idea. In early 1990 the consortium collapsed as it was unable to reach agreement about a proposed $1 billion chip-making plant. Each consortium member was asked to buy an equity stake in the venture and to agree to purchase a large amount of its chips. (So, the intent was to go into joint production after all! One wonders what the Department of Justice would have thought of that.) Apparently it was this latter requirement which derailed the entire effort. Members were just not willing to commit themselves to purchase from one particular source, be it even their own. With rapid technological change, the producers felt that they must retain the option to go after the most advanced components in the market, whoever makes them. To get ahead in today's competition it is more important to be nimble than big.

There are several lessons to be drawn from this failed effort. One is that vertical integration is too risky in the era of high technology. DRAMS obey other economic laws than railways or steel or oil. There are no considerable and indisputable economies of scale in the manufacture and marketing of semiconductors. It is not obvious that it helps to be very big. Another lesson is that the National Cooperative Research Act works the way it was intended to work. Consortia sometimes do a great job. Sometimes they break up. But they certainly do not breed monopolies. In low technology markets, there is an inherent tendency for a few sharks to swallow the small fish; eventually one victor and a monopoly emerges. That is why we need antitrust legislation. But high technology markets do not work that way. Cartels and monopolies cannot be maintained in a world of rapidly changing technologies. There is no need for society to seek safeguards against the abuse of market power in the high technology arena.

It is probably a mistake to view the race toward ever more powerful semiconductors as a battle of U.S. high technology versus the Japanese. The most promising course for a U.S. firm may be to seek out Japanese partners, rather than adversaries. Within a week of the collapse of U.S. Memories, Intel Corporation announced a deal with a Japanese chip company that will market the Japanese company's chips worldwide. In addition to three existing Japanese plants, a new U.S. factory is planned. And a few days later, IBM announced an agreement with the German Siemens AG to develop and market a new generation of 64-megabit chips.

Federal research laboratories. Perhaps the U.S. government should engage itself in technology development? Indeed, such activities do take place. The Department of Energy operates a series of user facilities, typically very large advanced scientific research laboratories that are made available to the scientific community. Examples are the National Synchrotron Light Source at Brookhaven National Laboratory, The Combustion Research Facility at Sandia National Laboratory at Albuquerque, the National Center for Electron Microscopy at Lawrence Berkeley Laboratory, the High Flux Isotope Reactor at Oak Ridge National Laboratory, and the Los Alamos National Laboratory.

Congress has required the Federal laboratories to direct a portion of their research budgets toward technology transfer. Every year, hundreds of separate technologies are transferred to existing firms and scores of new companies are formed based on technology spin-offs.

The Business of Incubation: Breeding and Nurturing New Companies

Business incubators can be operated by local government, spun off from academic institutions, or be run by private business. To incubate means to maintain controlled conditions favorable for hatching or developing. The business incubator houses and aids start-up companies. It provides manufacturing and office space and other amenities; it may offer secretarial support, computer services and management consulting services. Hopefully, after a few years, the tenants at the incubator will "graduate," moving out from the incubator and making it on their own.

States and communities seized on incubators as a job-creation vehicle in the early 1980s. Most incubators are nonprofit and receive financing from state or local agencies or from a university. It was expected that incubators would generate a steady flow of new companies, new tax receipts, and, most of all, new jobs. In particular, incubators were praised as the salvation of dying rural and urban economies. Incubators are often located in slums or in abandoned buildings because the operators can acquire the property cheaply or even free.

How efficient are the incubators? Obviously, the number of tenants is no measure of success, nor the number of "graduates." Managers of start-up companies, like any other management, naturally will be looking for low-cost operation alternatives. But will the incubator be able to nurture and graduate start-up companies that otherwise would not have made it? Many incubators have failed to live up to expectations either because they have played it safe by taking on as tenants only those businesses that were likely to succeed anyway, or because the graduates would relocate far away, providing no stimulus to the local economy.

A success story among the nation's incubators is the Austin Technology Incubator. It serves as a major outlet for the commercialization of technology developed at the University of Texas at Austin, and at the two major research consortia located in town: MCC and Sematech (see also the earlier chapter in this book entitled "New Webs and New Constituencies"). The incubator has been thriving as Austin has risen to become a major software center in the nation (many heavyweights in the computer industry including IBM, AMD, Motorola, and Dell have manufacturing and research facilities in town, and a host of minor producers of both hardware and software). In other words, much of the same mix that once produced Silicon Valley is currently driving the growth along the San Antonio - Austin corridor: a major national university, an industry park operated by the university, and strong synergy with a rapidly expanding business community. Significantly, there are now two daily nonstop airline connections between Austin and San Jose!

The conclusion seems to be that a business incubator can work wonders when it is able to serve as a regenerating center for private industry that is already in a growth mode. It is all a matter of synergy. A rapid flow of graduating incubator tenants will reinforce and will be reinforced by a thriving local economy. But it is doubtful that an incubator can do much to turn a stagnant or dying industry around.

Venture Capital

The start-up and the running of a high technology firm typically involves high risk. It is a "venture," a hazardous undertaking. The equity capital of the corporation may multiply one thousand fold. It may also be lost within a couple of months. The venture capital industry specializes in high-risk equity investments. Many venture capitalists (or venture capital companies, as the case may be) center their attention on emerging high technology corporations whose expansion is restricted by a lack of equity capital. With few assets and without a proven cash flow, such corporations are often unable to raise capital from conventional sources.

In a fashion, the modern venture capital industry acts the way the classical merchant bankers, the Rothschilds and the Morgans and the Mellons, used to do. It invests equity money in fledgling companies that it takes under its wings. The venture capitalist and the merchant banker can do this through the trick of a balanced portfolio where different risk elements hopefully offset each other so that the total risk content of the venture portfolio still remains within bounds.

Venture capital firms screen proposals based on four criteria: 1. The size of the investment. Most venture capitalists make individual investments in the range from $30,000 to $7,500,000. A study of some one hundred venture capitalists showed that their average portfolio size was ten ventures. Each capitalist cannot afford to spread his portfolio over too many small deals. 2. The technology and market sector of the venture. Venture capitalists must have some familiarity with the technology or the market of the proposed venture. 3. Geographic location of the venture. To maintain travel time and expense at manageable levels, many venture capitalists limit their investment activity to local ventures. 4. Stage of financing. Venture capitalists rarely invest seed capital and entrepreneurs typically have to turn to informal sources for this money. Startup capital refers to financing for establishing the operation; subsequent rounds of financing are used for expanding operations.

Just as the incubator wants its tenants eventually to "graduate," the venture capitalists wants to shepherd its clients to eventual disengagement and to free up its funds. It wants throughput. As the client corporation expands, the moment eventually will arrive when the venture capitalist helps the client draft a prospectus and to offer new equity stock on Wall Street (typically in the over-the-counter market). If the initial public offering (IPO) is successful (= sells at the projected stock price), the venture capitalist will cash out.

In 1979, Apple Computer had raised more than 7 million dollars from sixteen venture capital firms. On December 12, 1980 the company went public. The price of the stock increased so rapidly that the state of Massachusetts temporarily stopped people from buying it. By the end of the day, the company had reached the list of the Fortune 500. The stock market placed the value of Apple at $1.7 billion, which was larger than Ford Motor Company. Apple's rise had been meteoric, even by Silicon Valley standards. Steven Jobs was worth a quarter of a billion dollars. He was twenty-five years old.

But not all start-up companies make it. The venture capital industry is a risky line of business. Providing startup funding is always more risky than investing in the later stages of the development of a company.

There are also limited venture capital partnerships that private individuals can buy into. A partnership has a general partner, usually the venture capitalist's investment firm, which acts as manager. The limited partners -- the investors -- are passive. During the 1980s, investors poured more than $100 billion into partnerships which put the money into investments like apartment and office buildings, airplane leasing, oil wells, and cable television. But as many of these investments went sour, so did the partnerships. Prudential-Bache Securities Inc., once Wall Street's biggest purveyor of limited partnerships, eventually joined Sherson Lehman Hutton Inc. and other brokerage firms in pulling back from the troubled industry.

One of the most important sources of venture funds are existing companies buying a minority stake in a start-up company. Consider one of the luminaries of Silicon Valley: Sun Microsystems. SUN, headquartered in Mountain View, had been formed in 1982 to build computer workstations. It was quickly able to overrun the workstation market, outselling competitors such as Apollo and Hewlett Packard. Requiring vast additional funding, SUN in October of 1987 concluded a deal with AT&T. The agreement committed Ma Bell to buy up to 15 percent of SUN's stock over a three-year period. Analysts estimated the deal to be worth $300 million. It gave SUN the capital it needed to sustain its dizzying growth rate. In six years, it had cleared the $1 billion revenue hurdle. In its seventh year, SUN grew another 70 percent.

Small Business Investment Companies (SBICs) were created by an act of Congress in 1958. They are government-backed but privately owned. They represent a form of venture capital designed to help small businesses. An SBIC derives its initial capitalization from private sources; in addition it is eligible to obtain funds from the government through government-guaranteed loans. An

SBIC makes equity investments in small businesses; it can also make loans to its client company with a minimum maturity of five years. It is thus able to provide the client with both equity capital and borrowed capital.

Should only the private sector get involved in making venture capital available to prospective startups? Could a venture capital company be operated by central or local government, a federal laboratory, a state university, or even by a city? This is a difficult question that I shall return to later in this book (commenting on the role of industrial policy). If the venture company makes money, there is nothing to complain about, of course. But if it does not -- how then can one be sure that the public funds are not wasted? It would seem that there is very much a need for a venture capitalist to be able to go bankrupt -- to face the financial discipline that only a competitive market can provide.

The horror story that proves the point is Wedtech. Wedtech was a small company in the South Bronx that grew into a large defense contractor, riding high on corruption and fraud. The so-called Section 8(a) program, administered by the Small Business Administration, enabled a company to win federal contracts without subjecting itself to competitive bidding. The program was set up as an incubator for minority businesses. The general idea sounded innocuous enough: the client companies would be nursed along with government backing until they were large enough to graduate into the competitive marketplace.

The 8(a) program opened up fertile territory for cheating. Making fraudulent presentations to the SBA, Wedtech became certified as a minority-owned 8(a) contractor. Lining the pockets of officials of the SBA, of the Defense Department, and even at the White House, Wedtech was able to land a series of defense contracts, at prices way above Defense Department cost estimates. Submitting fraudulent filings with the Securities and Exchange Commission, Wedtech completed an initial public stock offering -- the first company in the history of the 8(a) program to sell its stock publicly. To investors, Wedtech presented itself as a booming enterprise, raking in government contracts and rapidly diversifying its business. The top officers were reported to earn more than $300,000 a year. The company operated plants on numerous sites in the South Bronx, in Michigan, and in Israel. To the SBA, the company portrayed itself as a struggling, unprofitable minority company, run by a disadvantaged Hispanic with meager income and holdings.

The Wedtech case exploded in 1986. The racketeering trial begun in 1988 eventually succeeded in dispatching some twenty-five Wedtechers to prison. Chastised for allowing Wedtech to rape and pillage the 8(a) program, Congress tightened the program rules, passing the Minority Business Development Reform Act. The new law requires competition for most federal contracts awarded by the SBA. But the problems did not go away. Also later, the 8(a) program has been plagued by instances of criminal misconduct.

Enterprise Zones

In the search for policy tools capable of stimulating the formation of new business, many state legislators and administrators have become intrigued by the concept of "enterprise zones." The idea is to use economic incentives to bring about neighborhood revitalization in depressed urban areas, and to foster economic development. Under various legislative arrangements, zone businesses get the benefit of wage tax credits (for employing residents of the zone), rehabilitation tax credits (for fixing up aging structures), tax deductions (for buying stock in enterprise zone businesses), and other tax benefits.

Enterprise zones were invented in the late 1970s in London to describe a plan to redevelop large industrial areas in the East End of that city. The idea was quickly imported to the U.S. More than 30 states have set up such zones.

In the wake of the 1992 riots in Los Angeles, much political controversy arose whether or not it would be a good idea to create federal enterprise zones. While most people agree that they would help to save our cities, many feel that the tax incentives should be combined with job training, education programs, and child care to help the residents in the zones.

Do enterprise zones work? Researchers have tried to measure their economic effects, but the evidence is ambiguous. It is easy enough to determine the lost tax revenue. It is more difficult to assess the indirect stimulus, locally and throughout the state. But perhaps it is wrong to measure the costs and benefits in monetary terms -- perhaps they should be measured instead in human terms, in terms of improved neighborhoods and new hope to inner city residents.

References

This chapter covers various institutional arrangements designed to promote the commercialization of new technology. This is the home turf of the IC^2 Institute, The University of Texas at Austin. For literature references, please turn in particular to refs [1],[3],[5],[6],[9],[10],[13].

"Silicon Valley, a thirty- by ten-mile strip..." J.L. Larsen and E.R. Rogers, "Silicon Valley: The Rise and Falling Off of Entrepreneurial Fever," in ref. [10].

"...consider the Case of Apple Computer." L. Butcher, *Accidental Millionaire: The Rise and Fall of Steve Jobs at Apple Computer*, Paragon House Publishers, New York 1988.

"Until recently, joint R&D ventures were considered..." See ref. [7].

"MCC was born in the wake of this announcement..." D.V. Gibson and E.M. Rogers, *R&D Collaboration on Trial*, Harvard Business School Press, Boston, Mass. 1994 and S.D. Stotesbery, "Improving Competitiveness Through Collaborative Research," in ref.[9].

"Business incubators can be operated by..." See ref. [2].

"It invests equity money in fledgling companies..." See ref. [1].

The account of the activities of venture capital firms is taken from T.T. Tyebjee and A.V. Bruno, "A Model of Venture Capitalist Investment Activity", *Management Science*, Vol. 30, September 1984, pp. 1051-1066.

"During the 1980s, investors poured more than $100 billion into partnerships..." J. Bettner, "Partnerships Investors are Angry -- and They are Suing," *The Wall Street Journal*, April 11, 1990.

"Consider one of the luminaries of Silicon Valley: SUN Microsystems..." M. Hall and J. Barry, *Sunburst: The Ascent of SUN Microsystems*, Contemporary Books Inc., Chicago 1990.

"To AT&T, the high tech alliance paid off handsomely..." It paid off in the sense that the SUN stock held by AT&T appreciated. But in the long run, SUN did not prove to be the partner that AT&T was looking for. Having instead bought NCR in 1991, it eventually sold most of its SUN stock.

"The horror story that proves my point is Wedtech." M.W. Thompson, Feeding the Beast: *How Wedtech became the Most Corrupt Little Company in America*, Charles Scribner's Sons, New York 1990.

"Do enterprise zones work?" R. Guskind, "Enterprise Zones: Do They Work?," *Journal of Housing*, Jan. and Feb. 1990, pp. 47-54.

CHAPTER 11

Riches From Junk

The spectacular rise of the junk bond in the 1980s, and the ascendancy of the investment banking firm of Drexel Burnham Lambert & Co. constituted a remarkable innovation in the financial arena. In 1990 the junk market nearly collapsed among defaults, bankruptcies, and recriminations. Two years later it rose again from the ashes, like a phoenix miraculously resurrected. Evolution sometimes occurs through the passage of chaotic disorder and crises. Today's junk market is very different from and much healthier than when Drexel dominated it. For one thing, there are more securities firms underwriting junk bonds, including many that also sell investment-grade bonds. The days are gone when one single investment firm could dominate both the supply of new issues and the placement of those issues in the market.

Background: The Market for Government Treasury Bonds

In order to understand what went so terribly wrong during the first euphoria, we need to appreciate the role of a "market maker" in financial markets. Let me begin with the market for U.S. Government Treasury bonds. During the course of its financing operations, the U.S. Treasury issues bonds (10-30 year IOUs), notes (12 months - 10 years) and Treasury bills (up to 12 months). The Treasury is the borrowing party. The ultimate investors are commercial banks, S&Ls, pension funds, private corporations, and even private individuals. In addition, there are middlemen in the markets, investment firms who act as brokers or market makers.

The market for U.S. government bonds is perhaps the best developed financial market in the world. It is the world's biggest and most important securities market. The Treasury bill, note and bond markets total about $2.3 trillion. The spread between the bid and ask price may for many maturities be as small as 2 cents on a bond with a nominal value of $100. Large blocks of bonds can be bought or sold within seconds with no noticeable effect on the market price. For all practical purposes, both the buyers and the sellers are price takers -- at any moment during the trading day the prices of bonds of various maturities can be considered as given and known. These parties in the market make decisions on purchases and sales based on the assumption of given prices.

For a market maker it is the other way around. A market maker in Treasury bonds sets prices at which he is willing to trade any maturity both ways -- a bid price at which he is willing to buy and an asked price at which he is willing to sell. Once these prices have been set, he passively executes trades as they arrive. The trades are made on his own account. As the pace of transactions rises and falls during the trading day, his own portfolio will wax or wane correspondingly. A market purchase of bonds will run down his inventory, a sale will replenish it.

If purchases in the market systematically tend to exceed sales (an excess demand at the current price is developing), the market maker needs to protect the dwindling holdings on his own account. He -- acting in unison with other market makers -- raises the asked price. If, on the other hand, a wave of selling develops in the market so that the inventory of the market maker balloons, he will adjust the (bid) price downward. In brief, while the buyers and the sellers make quantity decisions (the number of bonds to buy or sell), the market maker makes price decisions. As a result, an excess demand will cause the price to rise and an excess supply will cause it to fall. Equilibrium is restored.

Already in the late 1960s, the brokerage firm of Paine, Webber, Jackson, and Curtis put in place a computer program called Computrade, automating these simple rules. If the amount of securities the computer has been instructed to buy or sell at the quoted prices was exhausted, the computer would provide a new quote, up or down from the previous quote by a predetermined amount. A trader intervened to determine how much he could trade at the new price.

Corporate Bonds: Investment Grade and Below Investment Grade

Corporate bonds and municipal bonds are issued and introduced on the market through a process called underwriting. A financial firm such as an investment bank buys the bonds from the issuer (the borrower) and resells them at a higher price to the public, thereby assuming the market risk. That is, the underwriter is a market maker. If the issue is large, the investment banker may invite other houses to join with him in purchasing the issue. To facilitate the disposal of the securities he may form a selling group (a so called syndicate) for resale to the public. The investment banker will typically also serve as financial counsel and offer advice on the timing and terms of the new issue.

The Glass - Steagall Act of 1933 has until recently prevented commercial banks from underwriting securities -- both the underwriting of stocks and of bonds. The Act has maintained a clear line of division between underwriting and banking. I shall in a moment discuss recent moves by the Federal Reserve, granting underwriting powers to several large commercial banks, thus effectively blurring the line of division between market makers and investors.

Two well-known rating agencies, Standard & Poor's Corp. and Moody's Investors Service Co., rate the credit worthiness of bonds. The highest ratings, triple-A and double-A, are "high quality." The ratings single-A and triple-B are "medium quality." Double-B and lower is "below investment grade" = junk. The principal type of bond is a mortgage bond, which represents a claim on specified real capital, such as real estate, oil wells, plants. Another type is the collateral trust bond where the security consists of intangible property such as stocks and bonds held by the borrower and its ongoing business. In the event of financial difficulties befalling the borrower and subsequent reorganization, the holder of a mortgage bond will receive priority treatment (before the holders of subordinate debt).

The best measure of the risk of a bond at any given point in time is given by the spread between its current yield and the yield of a Treasury bond of comparable maturity. That is, the price of the bond reflects its riskiness. High quality corporate bonds typically yield about one percentage point more than Treasury issues of the same maturity; medium quality corporate bonds perhaps two points more. But these are broad benchmarks only; the pricing of a particular corporate issue will reflect both the general prospects in the industry where the corporation is operating and the performance of the corporation itself. For instance, traditionally bank bonds used to be considered as quite safe, reflecting the conservative stance and the sound finances of U.S. commercial banks. But times have changed, and as the losses of many U.S. banks mounted during the 1980s, investor confidence dwindled and prices of many bank bonds fell dramatically. Not even the largest and best known banks in the nation were immune against such fear. For instance, the yield of representative issues of Citicorp climbed in the fall of 1990 to some five percentage points above comparable Treasuries (the rating of Citicorp bonds at the time having been downgraded to single-A).

To a rational investor, investment grade bonds and junk bonds alike are perfectly valid outlets for investments. Corporations issuing investment grade bonds possess a high degree of creditworthiness, hence they need to offer the investor only a moderate yield. Issuers of junk have a lower credit worthiness; they must therefore promise a higher yield. How much higher? During the heydays of the junk market, junk yields were some 4.5 percentage points higher than Treasury yields. But as the market collapsed in early 1990, the spread shot up to an astonishing 12 percentage points.

These figures should be compared to the default rate on junk bonds. Drexel Burnham used to portray to clients the default rate as a bearable 3 % or so a year. According to Moody's Investors Service, Inc., however, the default rate climbed to 5.6 % in 1989 and to more than 7 % in 1990. Notice that the yield gap discussed a moment ago (between junk and Treasury bonds of comparable maturity) exceeded the default rate. The difference is the risk premium that had to be paid to an investor to induce him to choose the risky investment alternative rather than its actuarial risk free equivalent.

There has been some discussion how the default rate should properly be measured. Most junk bonds are still fairly recent issues (issued in the 1980s and the 1990s). The default risk calculated over the entire life span of these bonds is not yet known. The published rates were obtained by simply dividing total defaults by total outstanding debt. A more correct measure might be obtained by carrying out this division for each age class, just as actuaries calculate the death risk for individuals in each separate age bracket. Tentative calculations indicate that the default risk of junk has been much greater than commonly assumed.

Michael Milken: Financial Genius or a Criminal?

From his famous black X-shaped trading desk in Beverly Hills, Michael Milken propelled Drexel Burnham Lambert Inc. to its position as the leading underwriter in the junk bond market. Milken had joined Drexel back in 1969, quickly rising to become head of bond research. In 1978, he moved his newly formed high yield junk bond operation to California, to make use of a longer trading day. In 1983, total U.S. junk issues for the first time surpassed 5 billion dollars. Soon the figure exceeded 30 billion dollars. In 1987, Milken received from Drexel a stunning $550 million in compensation.

But Nemesis was waiting in the wings. In 1988, the Securities and Exchange Commission accused Milken and others of insider trading and other securities law violations. Drexel agreed to plead guilty to criminal charges. In 1990, Drexel went into Chapter 11 bankruptcy protection. Milken pleaded guilty to six felony counts related to securities fraud. As the year wore toward its close, he was sentenced to prison for violating federal securities laws but the sentence was later reduced. All in all, he spent 24 months in jail.

The unpardonable sin of Michael Milken was not that he engaged in some financial maneuvers of "stock parking." To the economist that would seem a rather minor offense. His sin was that he peddled to the investing public a greater amount of risk than it really was willing to hold.

The functions of a market maker are not limited to just buying and selling. They include the key function of assessing the creditworthiness of the issuer of the debt, so that the debt instrument is priced right in relation to its inherent credit risk (both price risk and default risk). But during those first explosive years the market for new issues never worked the way it was supposed to. The market was dominated by and controlled by Drexel. Drexel's share of the U.S. total never fell below 30 %. As one would later learn, Drexel was able to systematically overprice its new issues. Too muchy risky debt was dumped on the market -- more risk than the investors truly wanted to hold, had they been able to assess it properly. The subsequent downward price gyrations should be seen as convulsions occurring as the market tried to digest that excessive risk.

Leaving Michael Milken who is now back in society as a free man, and leaving also aside the engrossing saga of the rise and fall of the investment firm of Drexel Burnham and Lambert, let me turn to more enduring questions: is there a future for the junk market? Should there be a future for junk?

The Importance of High-Risk Debt

In order to attempt an answer, it may be helpful to state the obvious: Corporations need borrowed capital. They need debt. Banks and other financial intermediaries, and the financial markets, funnel the savings of private individuals (which as a group are the creditors in the nation) to corporations and to the government (the net debtors). The process of intermediation is useful to the savers, because it offers instruments of investment (like bank deposits) that are both liquid and safe. It is also useful to the borrowers, because it enables them to finance operations that they cannot fund out of equity alone.

Financial intermediation and financial markets constitute one of the great discoveries of Western civilization that once set capitalism into motion. The banking houses of Venice and Florence were great financial innovators of their times, providing funding for merchants, shipping, and early manufacturing. Eventually, financial intermediation was to blossom into the financial marvels of modern financial markets: Wall Street and the far-flung spectrum of modern financial assets.

It would be easy to argue that not only did the spectacular growth of financial intermediation accompany the maturing and development of the capitalist system, but that intermediation has actually been one of the driving forces of capitalism. The financial intermediaries of each age were with great adaptability and innovative power able to tap the available funds of savings and redirect and channel them to meet the credit needs of the day. This channeling function took the form of designing ever new and versatile forms of debt instruments.

So, there is nothing wrong with debt by itself. Perhaps, as innovation continues in the financial arena, one should see it as a natural process that the relative indebtedness of corporations continues its long run growth trend, simply mirroring the increased opportunities that financial innovations provide. But, having said this, clearly there also is such a thing as too much debt. Private individuals can suffer under the yoke of excessive debt. So can corporations. For any given corporation, given its line of production, its distribution and marketing, there will presumably exist some optimal mix of funding: short term debt, long term debt, and equity (net worth). Too much debt relative to equity is dangerous or undesirable for a number of reasons: 1. The equity of a corporation serves as a buffer to absorb temporary losses. If debt is excessive and equity low, a temporary lull in the cash flow which otherwise could easily have been absorbed by the company drawing down its liquid assets or stepping up its short term borrowing, can escalate into a major problem. The nightmare of the chief financial officer of every corporation is that he will not be able to meet the next debt payment (principal, or interest, or both). Sometimes it only takes one missed payment and the creditors will be howling at his doorstep. 2. Too much debt means that a corporation has already exhausted most of the potential credit that it could raise. If a favorable opportunity arises, such as exploiting the miscalculations of a competitor, management will not have the financial clout to make the necessary moves. When your banker hesitates, there is less maneuverability and less freedom of action. 3. Excessive debt is costly.

Most financial officers will tell you that the problem of excessive indebtedness is in the first instance one of an overhang of short term debt, and that if only their company were able to fund long term a greater portion of its total financial obligation (issuing mortgage loans, and bonds) their headache would go away. The onus of short term debt is that it has to be refinanced recurrently. Even if there is an understanding that the banker will automatically "roll over" the debt each time it expires, there is no guarantee that he will do so. The very moment a corporation runs into difficulties, he is liable to ask for additional collateral. A banker, as the saying goes, is a man who lends you an umbrella when the weather is fair and who wants it back when it starts raining.

One of the worst mistakes the management of a corporation can commit is to rely too heavily on short term debt. Long term assets and short term debt spell d-a-n-g-e-r.

Mortgage loans used to be the mainstay of corporate long term debt in this country. There are a number of historical reasons for this, all having to do with the fact that U.S. business eminence in the 19th and 20th century arose in industries that all happened to be endowed with plenty of natural resources or manmade real capital that could be mortgaged: farming, steel, oil, railroads, chemicals, and so on. U.S. wealth was built on natural resources, on machinery,and on plants. It was financed mortgaging those resources and those plants. The mighty house of Morgan arose, selling U.S. railway bonds in London. (Or, possibly, one could argue the other way around: precisely because these assets were mortgageable, economic growth was concentrated to these industries.)

In addition, established and well known corporations with healthy balance sheets, a strong cash flow, and good prospects can issue investment grade corporate bonds.

But what about all other corporations, those who have got a not quite prime cash flow, and who have already borrowed up to their mortgageable potential? From where are they to cover their additional long term funding needs?

The industries and the products of the future will not be based on natural resources, but on human inventiveness and human skills. Such qualities cannot easily be mortgaged. Human capital is accepted as collateral by no banker. How are the knowledge based industries of the future going to meet their long term financing needs? And where are new corporations, startups and initial public offerings on the stock exchanges going to find the long term debt without which they cannot exist?

I believe that junk bond financing will be crucial to U.S. industrial eminence in the next century. G. Gilder, a fellow at the Hudson Institute, points out that junk issues have been critical to the U.S. computer industry, providing between 1985 and 1989 some 80 % of all its finance. Timely junk issues enabled Compaq, Silicon Graphics, Cray Research and others to survive dangerous industry downturns. The American hard disk industry was saved by junk, including $250 million for Seagate. The crucial breakthrough for the fiber optics industry was MCI's fiber phone network, launched by William McGowan. Drexel provided no less than $3 billion of capital. Because of Milken's head start, the U.S. has laid about nine times as much fiber as Japan.

Seeing the advent of the "electronic superhighway" earlier than others, Michael Milken financed large building blocks of the electronic infrastructure that would become one of the major drivers of the U.S. economy in the 1990s. McCaw Cellular, which built the first national cellular network, raised several billions of dollars through Drexel. Turner Broadcasting and Tele-Communications Inc. (TCI) spearheaded the innovative drive in cable broadcasting, financed by multibillion dollar junk issues.

The collapse of Drexel was not in the interest of the economic well-being of this country. Drexel was sucked into a black hole of desperate long term financing needs of its customers. Milken improvised along the way, creating an innovative $200 billion new market virtually from scratch. The market grew too fast. Any capital market needs three parties: issuers (borrowers), market makers, and investors. Neither the market makers, nor the investors, had the personal and institutional and historical experience that it would have taken to handle this avalanche of loan demand wisely.

The responsibility of the financial advisor to the issuing party include the task of assessing the financial prospects of the customer in toto and advising on a sound debt strategy for the future. It includes an evaluation of the risk that the borrower will be assuming -- the cash drain of future debt payments, the lowered equity ratio and the narrowing financial maneuverability. It often includes the task of saying no.

Bankers know how to say no. The banker determines the credit worthiness of a corporate borrower and assigns a credit limit. In other words, bankers employ credit rationing. There is, as Keynes once said, a "fringe of unsatisfied borrowers". Once a borrower has hit his limit, it is of no use to ask for more of the same. (But the bank may be willing to discuss a new form of credit, such as following a first mortgage by a second mortgage, having its own terms and conditions.) There is an important lesson to be learnt here. Underwriting should be a stamp of approval -- that the underwriter has examined the borrower and vouches for the soundness of the issue. Not just for the present but also, as far as humanly possible, for the future, ten or fifteen years down the road. It should be a matter of trust. The entire financial world is built on trust.

I used to believe that there was great merit to the Glass - Steagall Act, barring banks from underwriting. It seemed to me to be wise to separate the underwriting and lending functions in the market. But I am not so sure any longer. The banks possess experience in assessing the creditworthiness of their customers that underwriters and merchant bankers do not have. Give banks underwriting powers and they shall be able to cater for all the funding needs of their customers, both short term debt and long term debt (underwriting corporate bonds) and even equity (underwriting the issue of new stock). Since 1989, the Federal Reserve Board has permitted a few large banks, including J.P. Morgan and Bankers Trust Company, to engage in underwriting. But the new powers come with restrictions, ensuring that banks segregate their underwriting business from their banking business. This means duplicating staff and limits on information-sharing among a bank's securities and lending officers.

The Future of Junk

Spurred by a climate of generally falling interest rates, the market for new junk issues staged a dramatic turnaround in 1991. Disgusted by the low rates offered on certificates on deposits and on investment-grade bonds, many investors took a new interest in junk. Also, corporate repurchases of junk bonds helped to buoy the market. Former Drexel clients such as RJR Nabisco and McCaw Cellular Communications seized the opportunity of buying in high-cost debt and retiring it. The repurchases shored up the companies' balance sheets and made their remaining junk bonds more creditworthy, causing bond prices to rise.

We may still be only in the beginning of the junk era. A number of current innovations could consolidate and strengthen the market, and even change its direction in the future. One straightforward development is the junk fund -- a bond fund specializing in high yield securities whose credit quality is rated below investment grade. Spreading its risks between different industries and different corporations, the bond fund presumably should be able to reduce the overall risk, while still offering the investing public an attractive return. Riding the wave of the resurgent junk market in 1991, the bond funds racked up an average total return of 36 % that year, counting dividends and price appreciation. And junk funds turned in two strong years in 1992 and 1993. Increasingly, high-yield debt is no longer offered to refinance existing junk or bank debt, but for expansion or other purposes. Some companies are even issuing junk to weather a rough patch of cyclically low earnings or industry overcapacity.

An even fancier innovation is the collateralized bond obligation, or CBO. A CBO is a pool of repackaged and over-collateralized junk. For instance, a $150 million CBO issue may be backed by junk bonds with a face value of $180 million. The diversification and extrapayments enable the CBO to receive an investment-grade rating from the credit-rating agencies. CBOs generally have three "tiers". The top tier, which consists of a pool of higher quality junk bonds, might pay an interest of about 8% to its holders. The second tier, somewhat less secure, might have a 11 % coupon rate. The bottom, and riskiest tier, may not pay a fixed interest rate. Instead, holders of third tier get the residual interest payments that are left once the top tiers have been paid.

Bibliographic Notes

Securities markets are usually analyzed in terms of portfolio choice. As explained by the winners of the Nobel Memorial Prize in Economic Science in 1990, H. Markowitz, W. Sharpe, and M. Miller, the key to portfolio choice is diversification. By diversification it is possible to reduce the overall risk content of a portfolio while still maintaining its yield. Portfolio choice, according to the Nobelists, consists of three steps: 1. Identifying the efficient diversified portfolio that minimizes the total risk, given any feasible expected yield. 2. Most investors are risk-averse, i.e. the investor must be offered a higher expected yield to offset any increased risk. 3. It follows that there will exist a particular optimal portfolio choice of each investor. The optimal choice involves the acceptance of some optimal risk.

Given the assessment of each investor of all risks involved, then, there exists an optimal portfolio choice for each investor, and a resulting market price. This is the so-called capital asset pricing model (CAPM). It assumes that all investor expectations are known and given. But what determines these expectations? In a famous chapter in his *General Theory of Employment, Interest and Money* (Macmillan, London 1936), J. M. Keynes likened the formation of expectations in the securities markets to a game of musical chairs where each participant is trying to outguess the others. Modern mathematical economists have described the speculative price "bubbles" that can arise from such collective action (on price bubbles, see Ref. [16]). Suddenly there is a rush to find the closest chair, a stampede to get out of the market. The market price nosedives. And perhaps there is something to be said for the notion that the collapse of the junk market in the late 1980s was indeed a case of a bursting "bubble."

The argument of the main text centers around the concept of information -- objective information about the prospects of the issuers of junk, enabling investors to form well-supported opinions about the investment risks and the attendant opportunities. If no objective information is available, the price formation in the market becomes perverse. To furnish information, investors need many information channels. They need a plurality of competing underwriters employing a plurality of market analysts. But if all information springs from one channel alone, be it a Michael Milken or some other guru, things are bound to go wrong.

References

"For a market maker it is the other way around..." U.S. Senate Joint Economic Committee, *A Study of the Dealer Market for Federal Government Securities*, U.S. Government Printing Office, Washington 1960 (written by A.H. Meltzer and G. von den Linde).

"The brokerage firm of Paine, Webber, Jackson, and Curtis..." *The Wall Street Journal*, Oct. 2, 1969.

"Tentative calculations indicate that the default risk of junk has been much greater than commonly assumed," E. I. Altman and S.A. Namacher, "Investing in Junk Bonds," New York University, and E.I. Altman, "Junk Bonds: A Whole New Ball Game," *Financial Analysts Journal*, May 1990. See also M. Winkler, "Junk Bonds are Taking their Lumps," *The Wall Street Journal*, April 14, 1989.

"From his famous black X-shaped trading desk..." J.B. Stewart, *Den of Thieves*, Simon & Schuster, New York 1991.

"G. Gilder, a fellow at the Hudson Institute, points out that ..." G. Gilder, "The Drexel Era," February 16, 1990, "The War Against Wealth," September 27, 1990, and "America's Best Infrastructure Program", March 2, 1993, all in *The Wall Street Journal*.

"There is, as Keynes once said, a 'fringe of unsatisfied borrowers'..." For this reference, as others relating to the so called "availability doctrine" in monetary policy, see A. Lindbeck, *The 'New' Theory of Credit Control in the United States*, Almqvist and Wiksell, Stockholm 1962.

CHAPTER 12

Thrifts and Banks in Turmoil

If there ever were a case for an economic theory of chaos and turmoil, it would seem that the U.S. banking system and the savings and loan industry are the prime candidates. Rocked by corruption, scandals, and bancruptcies in the 1980s and early 1990s, these financial institutions were brought to the brink of the abyss.

The features of evolution of the financial system are the same that we have already examined in the production sector of the economy. First, diversity: Almost every month, new species of financial assets and financial debt are launched in the marketplace. The reader who has been through the gruelling process of refinancing a piece of real estate will know what I am talking about -- the menu of available forms of fixed-rate and variable-rate mortgage financing is bewildering. There seems to be no end to the ingenuity and creativity in the financial world, generating a continuous stream of financial innovations.

Second, complexity: A certificate of deposit, mortgage-backed securities, zero coupon bonds, mutual fund junk bonds are all examples of high tech financial products. Financial high tech involves new types of complex financial transactions, or even entirely new financial institutions like credit card companies, money market funds, or Euro banks.

To understand what a financial technology is, consider for a moment the entire financial system in a nation. It acts as an intermediary between savers and borrowers. On occasions, a willing saver may get in direct touch with a willing borrower and they may arrange a financial deal directly between them. This happens when a nice uncle or aunt provides a loan to a needy niece. It also happens when the owner of a private residence wants to sell the property and is willing to extend "private financing" to the buyer. But usually the funds of the saver are channelled through one or several financial intermediaries before they reach the ultimate borrower. A farmer in Missouri saves in a local savings and loan institution. The deposit is channeled, via the market for brokered deposits, to a large bank in New York. The bank buys bonds issued by a development bank in Brazil. The development bank lends money to a Brazilian real estate project. In this manner the chains of the financial logistic system span the world.

The financial world performs the miracle of converting comparatively safe funds into riskwilling funds. Financial institutions issue deposits that provide a high degree of safety (low risk) to the investor. The institution will set aside some liquid reserves; another portion of the funds may be placed in government securities and other investment outlets that are safe and yet yield a satisfactory rate of return. The remainder is lent to loan-seeking corporations and individuals. So, some fraction of the deposits is converted into riskwilling funds. I should hasten to say that many banks and S&L's in the U.S. have gone

Milestones in Modern Banking	
1956	Bank Holding Company Act, allowing banks to circumvent state branching laws and permitting holding companies to engage in mortgage banking, finance company operations and factoring.
1961	Negotiable certificate of deposit (CD), introduced by National City Bank of New York.
1966	Eurodollar CDs, introduced by Citybank through its London office (a Eurodollar is a nonresident dollar redeposited abroad).
1980	Depository Institutions Deregulation and Monetary Control Act.
1982	Garn- St. Germain Act Depository Institutions Act.
1982	CHIPS (Clearing House Interbank System) in NewYork, an electronic system for clearing Eurodollars.
1983	SWIFT(Society for Worldwide Interbank Financial Telecommunications), an international computer network.
1984	Federal Reserve Board automatic clearinghouse (ACH), an electronic network for clearing of member bank funds.
1989	President Bush signs the savings and loan rescue bill creating the Resolution Trust Co. Banks are permitted to own savings and loan institutions.
1989	The Basel Accords implemented, under the auspices of the Bank for International Settlements in Basel. Bank supervision throughout the world is coordinated, requiring banks to maintain capital equal to 7.25% of business and most consumer loans (requirement raised to 8 % on Jan.1, 1993).
1991	In November, Congress passes new legislation strengthening bank regulation.

overboard in their willingness to accept lending risk. Something will be said about this later; at the moment I just want to point out that the role of creating riskwilling capital is part of a functioning financial system.

Following the financial logistics network downstream, the lending product acquires also other attributes than riskwillingness, such as easy interest rates and attractive, payback schedules, and flexibility in terms of the requirement for collateral.

I have pointed to the diversity and the complexity of financial technologies. Third, and as a result of the drive toward ever increasing diversity and complexity, there is self-organization and evolution. Most forms of financial assets and debt, and most financial institutions, arose spontaneously, in the free and unregulated marketplace. Some were created by the government (notably the issuance of government money and the creation of the Federal Reserve System signed into law by Woodrow Wilson in 1913).

Within the confines set by government supervision and regulation, financial institutions in this country have for some time been engaged in rapid evolution. Some of this evolution has been prompted by the advent of new information technology, computers, and electronic money. But, unfortunately, much of this dynamism has been directly prompted by the government itself, through unwise regulatory decisions. The excesses were exacerbated by incompetent financial managers.

The Savings and Loan Debacle

According to the old order of things, there were two kinds of financial institutions: banks and thrifts (such as S&Ls and savings banks). There was a division of functions between the two. Banks accepted demand deposits that paid no interest and were payable upon demand, and provided medium term loans to manufacturing and local industry. Thrifts accepted time deposits and provided 15-30 year real estate loans. The little neighborhood savings and loan was immortalized in the classic Frank Capra film It's a Wonderful Life. In the film, Jimmy Stewart played the director of a sleepy hometown thrift.

But with the Depository Institutions Deregulation and Monetary Control Act of 1980, the border line between banks and thrifts became blurred. Both banks and thrifts were permitted to pay interest on so-called NOW accounts (=negotiable order of withdrawal). Thrifts starting competing head on with banks, providing short-term and intermediate term financing also outside the real estate business which was the field of their expertise. Both kinds of institutions solicited deposits, on very similar terms. They both provided checking facilities.

The negotiable certificate of deposit (CD) is an interesting innovation that offers the saver an alternative to the more old-fashioned time deposit. The CD has a fixed maturity, such as 3 months, half a year, or one year. The financial institution is able to plan its own cash flow, investing the funds longer term and more profitably than if the funds could be withdrawn at will. The financial institution can therefore offer a higher rate than on ordinary time deposits. Still, from the point of view of the saver, a bank CD is a very liquid form of savings, because the CD can be sold in the secondary market before maturity.

For many years the so-called regulation Q set a ceiling on the rate that financial institutions could pay on time deposits. The Depository Institutions Deregulation and Monetary Control Act called for a phasing out of regulation Q. With the interest rate caps gone, the institutions felt free to enter into a costly courting of the favors of the depositors, offering them free barbecue grills, free trips to Las Vegas, and outlandish interest rates which the institutions could not afford. A new category of financial intermediaries made their entry onto the financial arena: the deposit brokers. They scour the nation each morning for the highest interest rates being paid that day on CDs, and then purchase such deposits in multiples of $100,000 -- the limit on each deposit insured by the federal deposit insurance. When the certificates mature the money would again flow to whomever was paying the best rate.

The S&L industry was caught in a vicious circle in the 1980s that set the thrifts on their course to destruction. Double-digit interest rates in the early years of the decade had caught thrifts between low fixed-rate loan rates and skyrocketing deposit costs. The dilemma was exacerbated by institutions rushing to jack up the interest rates paid to depositors in vain attempts to attract more savings. To desperate managers, there seemed to be only one way to enhance the sagging profitability of their institutions: high-risk, high-return loans. And so they loaded up on short-term and medium-term loans to business, oil, and real estate.

The Garn- St. Germain Act of 1982 went far toward the deregulation of the thrift industry. One single shareholder could now own a savings and loan. There was no need to rely on local savings; funds could be brought in as desired in the form of brokered deposits. The act no longer limited operations to lending; direct investments in condominiums, shopping malls, or even in the movie industry might seem more attractive. In effect, many S&Ls came to function as a kind of investment banks. American Diversified in California operated from the ninth floor of an office building, had no tellers, did no consumer business and never wrote a home mortgage loan. With about half a million dollars in equity, it grew to $1.1 billion before closed down.

In retrospect it is clear that the deregulation of the thrift industry in the early 1980s was a tragic mistake. There is nothing wrong with financing risky projects. But the officers and owners of the thrifts never bore the risk themselves. The risk was borne by the government, and, in the last instance, by the taxpayers.

The savings and loan bailout bill signed by President Bush in August, 1989 created a new federal agency, the Resolution Trust Co. The bailout came to cost more than $200 billion in bad loans. As the Resolution Trust Co. reorganized the failed thrifts and started selling the viable portions of their portfolios, it ended up sitting on vast holdings of assets that had gone sour: junk bonds, real estate, and direct investments. At one point, the Resolution Trust Co. was the largest single holder of real estate in the land, a socialized property-owning monster. Having sold the last of the defunct S&L's under its dominion in 1995, the RTC retired into obscurity.

Under the new law, the S&L's must keep nearly 70 % of their assets in mortgage-related investments. Thrifts are forced to employ the more realistic accounting practices of commercial banks and are subject to banks' tougher capital standards. The Federal insurance for S&L's has been taken over by the corresponding agency for the banks (the FDIC, the Federal Deposit Insurance Corporation).

What will now happen to American thrifts? Will they survive as separate entities, or will they disappear, merging with the banking industry? Does the future high tech economy need S&L's?

One thing is certain. The economy needs long-term savings capital, both now and in the future. The U.S. has one of the lowest savings ratios of Western nations. The gross savings ratio during 1979-85 was a little less than 18%. The corresponding figure for West Germany was 24 % and for Japan 32%.

But there are other forms of long-term financial saving emerging. The foremost of these newcomers is the money market fund -- a mutual fund which invests most of its resources in Treasury bills and short-term Treasury bonds. Money market funds do not need to hold any reserves as banks and thrifts do, and can offer rates of return which are quite close to the current money market rates. Some of these funds are huge and have established brand names and a solid track record of safety and sustained returns. The fees are often low (most funds are open end, i.e. they issue new shares as required, there are no "loads" upon entry or exit, and the management fee runs to half a percentage point or less).

Financial Innovation: Packaging and Re-packaging Mortgage Securities

The development of secondary loan markets for residential mortgage has revolutionized residential financing in the U.S. If a private person buys a home today and applies for a mortgage loan from a mortgage corporation or a S&L or a bank, chances are that within a few weeks that originator of the loan will have sold it to some large portfolio management institution operating nationwide and located far away. The ultimate investor has no personal relationship with the customer. The customer will receive in the mail a notice about the new address to where monthly payments should be sent. The ultimate investor does not need any direct personal contact with the borrower, once the terms of the loan has been set and the risk category of the borrower has been assessed (by the originator of the loan). From then on, the home owner is just a piece of statistics and the law of large numbers makes it possible to calculate with great precision the expected return and the expected cash flow to the ultimate lender.

As a result of the financial collapse of the Great Depression, the Federal Housing Administration was created to insure lenders against the default of mortgage loans, and the Federal National Mortgage Association ("Fannie Mae"), whose function was to purchase mortgages. Fanny Mae was reorganized as a privately owned corporation in 1968. Those programs requiring government subsidies or other direct Federal support were assumed by the Government National Mortgage Association ("Ginnie Mae").

So-called mortgage-backed securities were invented by Ginnie Mae in 1970. Such securities are issued by private mortgage institutions. The Government guarantees the timely payment of principal and interest. They are backed by pools of Government-insured or -guaranteed mortgages. The securities thus represent ownership in the monthly payments from a pool of mortgages. The originators of the mortgages retain responsibility for servicing the mortgages and "passing through" the homeowners' payments to the owner of the security, each of whom receives a *pro rata* share of the monthly payments made on the mortgages in the pool.

The success of the Ginnie Mae program encouraged private mortgage originators to issue mortgage-backed bonds without government involvement. Such securities are in many respects similar to other corporate bonds. They are obligations of the issuer with a stated maturity and fixed interest payments. They are collateralized by pools of mortgages. They enable thrift institutions to borrow against their mortgage assets to obtain funds for new loans.

In the early 1980s the mortgage securities market exploded when Salomon Brothers Inc. and First Boston started repackaging mortgages on their own and selling mortgage bonds. The market grew to fill the gap left by the departing thrift industry, and grew even more as financial innovation occurred ("collateralized mortgage obligations" and "real estate mortgage investment conduits"). The next milestone: the Resolution Trust Company -- the 800 pound gorilla thrift cleanup agency -- plunged into the mortgage-backed securities market. Issuing its "Ritzy Maes" as Wall Street calls them, backed by mortgages from busted thrifts, the RTC swept billions of dollars worth of such mortgages off its books.

In sum, rapid innovation has made the conventional thrift industry obsolete. Such innovation has occurred both as new more competitive conduits for savings have appeared (money market funds) and more competitive ways of packaging and selling mortgages have been invented.

Electronic Money

The nature of commercial banking has changed dramatically in the U.S. during the last 35 years. Banks supply the general public with a means of payment -- money. Money used to be notes and coins and demand deposits. But now it is also "plastic money." We are on our way to the "cashless society."

It all began with the automated teller machines (ATMs). Using language that I have introduced in an earlier chapter, the teller machine is a robot. It consists of software and machinery for dispensing dollar bills. You stick in your card and enter your password. The teller machine checks the password and clears you for transactions (feedback). You issue an instruction to withdraw forty dollars. The machine spits out two twenty dollar bills.

But it is when the teller machine is placed in a supermarket or a gas station that things become really interesting. You enter your password and the amount to be paid, and the funds are electronically debited to your account and credited to the store. (Americans have yet to wait a few years to see the full system, but it is common in Europe.) Electronic money takes the place of conventional money. There is no need to write a check, nor to pay in cash. The electronic money is as good as cash. Actually, it is better, because the owner of the store does not need to walk to the bank at night to deposit it. It has been deposited there already, instantaneously.

This will have important effects on bank balance sheets. There is something that bankers call "float." It consists of checks that you and I have written and that are still in the mail or under collection but which have not yet reached the bank. They are still in the pipeline. They constitute a large collective debit item. Float artificially inflates the book value of demand deposits. But to banks, float is good money that can be invested or lent. Electronic funds transfers is instantaneous, without float. Banks then lose funds that would otherwise have stayed in their coffers for two or three days. It is not uncommon for the float of a bank to amount to 10 % of its deposits. So, banks are not too keen on electronic transfers. The real cost to them of such systems is not the installation of hardware. It is the effect on their gross deposits.

Most banks issue credit cards (Visa or Master Charge). A credit card has functions beyond a debit card, because the outstanding balance need not be paid off in full. Actually, the bank will prefer that you do not. The outstanding balance will appear in the balance sheet of the bank as an unsecured consumer loan. It is, from the point of view of the bank, a high-yielding but risky investment of funds.

Credit cards have been around for a long time. When the wives and daughters of wealthy English landowners in the last century travelled to London for shopping or social functions, they would simply leave the gentleman's calling card in the shop or at the club. Bills would be presented for payment later. In 1915 Western Union began issuing credit cards to preferred customers. The modern credit card was created in 1950 by Diners Club.

The exorbitant rates that you have to pay on your credit cards is an invention of a more recent generation of bankers. Some credit card companies will charge you 18 % per annum or more on your outstanding balance. Defending their practices, the companies point to their large losses. Yes, a credit card company will surely rack up losses commensurate with the risk that it assumes. But the risk itself can and should be controlled. Remember that application form that you filled out before you got your credit card? Already in the early 1960s operations research analysts showed how it is possible to design a questionnaire and a system of points to be used to grade the answers, in order to construct a statistical universe of accepted applicants with a known and given default risk. An applicant scoring above the cutoff grade point rating, is accepted. An applicant that scores below it is turned down. It is all a matter of mathematics and probability theory. One of my own students built a billion dollar consumer credit company using those very same principles. But the credit card managers of today do not seem to want to control risk. They mail out cards and sign up applicants by the millions, crossing their fingers and hoping that everything will go well.

I find it astounding that the credit card companies have not eyed the obvious opportunity to market a high-grade card to a select clientele, charging only a moderate interest rate per annum. That may have been the original purpose of the "premium" or "gold" card. Using credit payments statistics from a regular card, it would be a simple mathematical exercise to develop a list of criteria that select a sub-population of cardholders with any desired low credit risk. But the temptation to mass distribution is irresistible, I guess. And there are enough fools among the general public willing to pay a premium annual fee for the imaginary prestige of the color of gold, embossed on a piece of plastic.

The new memory card developed by the French contains a microprocessor or similar technological device which can provide information, memory, and even logical processing facilities.

The next step is home banking -- accessing your bank account directly from a personal computer sitting in your own home. It would work just the way you can access your stock brokerage account today to order purchases and sales of common stock.

It is not obvious that home banking will be a blessing to banks. What if national money market funds would offer the same services? There are already money market funds that offer check writing facilities. Suppose that you could access your money market account in New York or Chicago, and electronically effect payments to your local grocery store. Who would then need a bank account?

The Sorry Story of U.S. Banking

Let me turn to bank lending. The future looks ominous here too. Banks are taking too great lending risks, pushing flashy innovations but forgetting basic principles of prudence.

Banks used to rely mainly upon demand deposits for their resources. A bank would loan less than half of its deposits; of the remainder a large chunk would be invested in government bonds. A senior banking official from those days would look in horror at the balance sheet of a bank of today, perhaps asking if the bank was ready to go into receivership. Well, some of them are. An entire generation of more aggressive bankers followed, honing the skills of active "liabilities management," always standing ready to raise funds as required, either in the money markets or in the CD markets and being willing to go after high yield loan projects (with commensurate high risk) that earlier bankers would never have touched.

The sorry track record of banks in the early 1980s demonstrates that bank managers grossly and recklessly have underestimated the need for liquid reserves. During the time period 1984-1991, more than 1,100 banks went belly-up. The first task of a financial institution is to stay sound. Solvency is the number one priority. Profitability must come much later. Admittedly, it is not easy to assess the risk. It is one thing to assess the potential of one single manufacturing project, say. It is quite a different matter to evaluate the *industry risk*, the risk that an entire sector of the country runs into trouble, like the oil industry or the computer industry. It just is not enough for a banker to carry reserves as a protection against individual loan risk. A banker also needs reserves against industry risk. How do you assess such risk? Well, what most bankers do and what the bank examiners also do, is to look at other banks. A bank is deemed prudent if it keeps reserves that look good compared to what other banks of similar size are doing. But these subjective standards have been eased over the years. The banks have indulged in an act of collective grandiose self-deception, lowering their holdings of reserves and liquid buffers of government bonds.

I am amazed when I read or listen to bank advertisements. Infallibly, the customers are offered more attractive deposit arrangements and higher yields. Why do not banks advertise that they are safe? Of course the marketing people and the admen know what they are doing. The fact must be that the general public does not care. If we do not demand safety, we will not get it. If we condone sloppy portfolio management, we will get it. If we are willing to "forgive" (just as bankers are asked to "forgive" defunct borrowers), then the bankers will certainly exploit us.

One of the great mysteries of Saving is that ordinary men and women who work so hard to earn their dollars, turn into naive sheep when it comes to investing. It has been the same story over the centuries. Borrowers take the savers' money and run. The government issues bonds and debases them through inflation. Bankers solicit our deposits to pour into their pet projects. And when the bank goes under, the FDIC (the Federal Deposit Insurance Corporation) moves in late at night, with new stationary and new promises. Next morning it is business as usual, under another name. The money is lost, but who cares?

There is only one solution to this problem. Society must protect itself. It does so in the form of bank legislation, bank regulation, and bank inspection. The new banking legislation passed in November 1991 establishes a new regulatory regime by firmly linking supervision to bank capital. Three supervisory agencies (the Federal Reserve Board, the Comptroller of the Currency, and the Federal Deposit Insurance Corporation) all issue guidelines for the evaluation of bank capital adequacy. The new legislation tightens these capital requirements substantially. It forces the regulators to pull the trigger much sooner on sick banks. If capital declines, the regulators are required to take corrective action such as reductions or suspensions of dividends, stock sales to raise capital, and management changes. Banks may be closed if their ratio of capital to assets falls to 2 %. New bank examiners are being trained, and their pay is being upgraded. Regulators are required to perform annual, on-site bank examinations.

The new banking law also raises the premiums that banks have to pay for federal deposit insurance. And, to bolster its shrinking insurance fund, the Federal Deposit Insurance Corporation was granted a §70 billion dollar loan, to be repaid by future premiums.

The recession of 1990-1991 hit U.S. banks severely. Large sectors of the U.S. economy were ailing. Sound and promising lending projects were hard to come by. It became difficult to lend money wisely. At the same time, the drive of federal regulators for a stronger capital base was unrelenting. The solution for many bankers was to curtail their lending drastically and to pour the trickle of new deposits into government bonds instead. A "credit crunch" developed, exacerbating the shrinking economy. The Bank of New England in Boston went under. Other banks merged in a frantic attempt to restructure. Chemical Banking Corp. acquired Manufacturers Hanover, NCNB Corp. acquired C&S Sovran, and BankAmerica acquired Security Pacific. These are but the latest examples of the ongoing shakeout in the banking industry. Nobody knows what the industry will look like when the dust eventually settles.

To put matters in a nutshell, the long-term problem of banking is that banks no longer seem to be able to offer a unique product where they have a clear competitive advantage. The checking facility offered by banks used to be such a unique product. With the advent of electronic and plastic money, checks face stiff competition. On the lending side, banks would lend primarily on the strength of a company's cash flow statement. But in so doing, they neglected the most dynamic segment of the lending market: asset-based lending, that is, loans secured by accounts receivable, inventory, and business equipment. This portion of the market has instead been picked up by to the so-called commercial finance companies, which now are responsible for a full one third of all commercial and industrial lending.

There is a growing secondary market for business loans with standardized collateral. Once having signed up the loan for the corporate customer, the local banker may decide to get rid of the risk, selling it (at a discount) in the secondary market. Also, once this secondary market gets established, there is no reason why a loan should be initiated by a commercial bank. I can see an entire generation of new specialized corporate financing companies initiating such loans and then reselling them in the secondary market. To discuss the possibility of such a loan, a corporation would get in touch with the local office of the financing company rather than a bank. There could be such financing companies specialized in financing warehouses, machinery, and plants in various industries. The financing companies would either sell the loans outright in the secondary market or they would package and bundle them and use them as collateral for securities.

Commercial banking is one of the great achievements of human civilization. Its evolution can be traced from the first "bancas" of the money-changers in Venice (banca = table), via the Medici family in Florence, and the financial empire of Jakob Fugger in Augsburg spanning early 16th century Europe. Gradually, in the process of time, a code of professional ethics developed. It stressed the honorable reputation and the good name of the bank.

Banking in the United States has had its ups and downs. Its history has been peculiarly checkered with panics, failures, and runs. The go-go years of the 1980s and the ensuing bust bear a disquieting resemblance to the boom-and-bust cycle of the 1920s, culminating in the stock market crash in 1929 and the banking "holidays" (= closings) in 1933. That time, the banking crisis prompted Congress to put safeguards in place like caps on savings interest rates (to dampen the excess of the boom) and federal deposit insurance (to alleviate the damage wrought by the bust).

Why is it that bankers in this country always play for such high stakes? One year it is the lure of South American loans, another year the calling of venture financing. The step from the sober financier to the aggressive huckster has never been long. Is it something in the national character? The Swiss are a phlegmatic and conservative lot. They make the world's best bankers. Americans are entrepreneurial and gambling and risk-willing. That makes for outstanding innovators and founders of cutting-edge companies. But it makes poor bankers. No -- correction: it makes dangerous bankers that need to be reined in.

What this country needs is not banking reform but a reform of bankers -- a return to the basics of banking and a new attitude of bankers to their profession. Bankers need to rediscover their customers, that they are people in flesh and blood and not just numbers on a computer printout. Bankers need to discover their corporate customers, visit with them, talk to them, and get an understanding of their banking needs. A new breed of bankers is needed in the high-tech world: bankers that are able to communicate in a meaningful manner with the manager of a modern hospital, of a software company, or of a start-up biotechnology firm.

Bankers need to be recruited and trained in new ways to meet these challenges. Perhaps the banker of the future should have a background in medicine. Or in botany. The primary concern of bankers in the high tech world must be to build enduring customer relationships. To illustrate the kind of close cooperation that is needed, I think bankers should participate, on a regular basis, in the product development decisions of their customers. Companies need financial commitments by their bank that span the entire life cycles of their products, right from the moment the product development decision is made. The bank, on its side, needs an ongoing relationship in order to be able to assess the credit risk fairly and to be able to design a financial package that caters for the evolving needs of the customer.

Will the bankers see the writing on the wall? I can see two alternative future scenarios -- a bifurcation, as it were, of the financial system in this country arising out of the present chaos. The first alternative is that the bankers somehow will get their act together. The second is that the historical development will simply bypass the banks, leaving them behind, as remnants of an earlier financial order that failed. Instead, rapid financial innovation will thrust a new breed of financial institutions onto the arena left by the departing banking industry.

Bibliographic Notes

It is becoming commonplace to consider the excesses of S&L executives and bankers during the 1980s as a temporary deviation, a "decade of greed," a speculative bubble. The aim of my text is to look deeper, examining how evolving innovations in the financial arena actually created opportunities for more effective intermediation and for better risk management, but how these opportunities were thwarted by an increasing readiness by the managers of the financial industry to assume greater portfolio risks. And, of course, this shift toward a greater risk exposure took place inside the setting of a reckless experiment of financial deregulation conducted during the decade.

The flow of funds in the entire financial sector of the economy may be visualized as a flow-of-funds network, see S. Thore, "Credit Networks," *Economica*, Feb. 1969, pp. 42-57, S.Thore and E.Vardal, "A Programming Model for National Credit Budgeting in Norway," *Economie Applique*, 1974, pp. 423-456 and several additional publications dating from the early 1970s culminating in the comprehensive study S. Thore, *Programming the Network of Financial Intermediation*, Universitetsforlaget, Oslo 1980. Financial innovation is represented as new nodes and new links in the network. Each link represents a financial "activity" (in the sense of activity analysis) or, to use another term, a financial technology. "High tech" in the financial world, then, alludes to the lengthening and proliferation of the intermediation chains, and the gradual refinement of the vector of attributes possessed by the final credit product, to fit the varied demands and requirements of the borrowers.

The recent pace of financial innovation has been breath-taking. It is still too early to trace the outlines of the resulting structural change of the nation's flow of funds. Perhaps the reader will feel that my attitude to the S&L's and the banks is too cynical and too pessimistic. Recall, however, that few of the underlying problems of the turbulence have yet been addressed. Many financial institutions are as vulnerable as ever. Deposit insurance still props up too many small and inefficient banks. Bankers do not yet seem to have got the message that the corporations of the future are going to be small and hold preciously few assets that can be mortgaged -- that these corporations will need credit based on the promise of their technologies and their cash flow, not on their holdings of warehouses, machinery, and real estate.

References

"But with the Depository Institutions Deregulation and Monetary Control Act of 1980..." E. Brewer and Others, "The Depository Institutions Deregulation and Monetary Control Act of 1980," *Economic Perspectives*, Federal Reserve Bank of Chicago, September/October 1980, pp. 3-23.

"The Gam- St. Germain Depository Institutions Act of 1982..." G. Garcia et al., "The Gam- St. Germain Depository Institutions Act of 1982," *Economic Perspectives*, Federal Reserve Bank of Chicago, March/April 1983, pp. 3-31.

"American Diversified in California operated from the ninth floor..." L. White, *The S&L Debacle*, Oxford Univ. Press, New York 1991.

"...the deregulation of the thrift industry ... was a tragic mistake" E. Kane, *The S&L Insurance Mess: How Did it Happen?* The Urban Institute Press, Washington D.C. 1989 . See also L.G. Sandberg, "The U.S. S&L debacle -- was the hazard moral?," *Skandinaviska Enskilda Banken Quarterly Review*, No.3-4, 1992.

"The bailout will cost -- " The bailout bill requires the RTC to wind up by 1996. In March 1993 the Clinton administration sought $45 billion to wrap up the bailout effort. If the entire amount is used, it would bring the total tab for the bailout to $194 billion. See also M. Mayer, *The Greatest Ever Bank Robbery: The Collapse of the Savings and Loan Industry*, Scribner's, New York 1992.

"So-called mortgage-backed securities were invented by Ginnie Mae..." C.M. Sivesind, "Mortgage-backed Securities: The Revolution in Real Estate Finance," *Federal Reserve Bank of New York Quarterly Review*, Autumn 1979, Vol.4, No.3, pp. 1-10.

"The nature of commercial banking has changed dramatically ..." J.A. Haslem, *Bank Funds Management: Text and Readings*, Reston Publishing Co., Reston, Virginia 1984 and D.B. Graddy, A.H. Spencer and W.H. Brunsen, *Commercial Banking and the Financial Services Industry*, Reston Publishing Co., Reston, Virginia 1985.

"But it is when the teller machine is placed in a supermarket..." S. Thore and I. Eriksen, "Payment Clearing Networks," *Swedish Journal of Economics*, Vol.75, 1973, pp. 143-163.

"Already in the early 1960s, operations research analysts showed how..." See several papers reprinted in Part III ("Lending and Credit Functions") of *Analytical Methods in Banking*, ed. by K.C. Cohen and F.S. Hammer, Irwin, Homewood, Ill. 1966.

"An entire generation of more aggressive banks followed ..." J. Beebe, "A Perspective on Liability Management and Bank Risk," *Economic Review*, Federal Reserve Bank of San Francisco, Winter 1977, pp. 12-25.

CHAPTER 13

The Social Management Of Technology: Where Are We Headed?

For centuries enlightened people have understood that technology changes the society that we live in. Historians, social thinkers, and psychologists have explored the matter, trying to analyze how technology impacts on the conduct of war and political history, on human civilization, on customs and mores, on life on the farm and in the city, on man's perception of himself and the world around him, even on his artistic expressions.

The Blessings and Curses of Technology

Early thinkers tended to stress the beneficial effects of new technology. Nobody could reasonably dispute the blessing to mankind that Gutenberg brought in the form of the printed book. Improved methods of navigation and advances in cartography spurred the great wave of geographical exploration in the 16th and 17th century. Merchant banking and new techniques of financial intermediation provided the financial background for these undertakings. Early medical advances like variolation against smallpox relieved human suffering. Catherine the Great brought in an English doctor who successfully inoculated the empress with matter from the pustules of a convalescent young boy and 140 aristocrats immediately followed her example. Variolation clinics were set up in several provincial cities; before the end of the eighteenth century some two million Russians had been inoculated.

Technological progress has of course always had a darker side: the development of new and more deadly military technique -- the chariot, the catapult, armor, the cannon, the rifle, smokeless powder, the Big Bertha (the 98-ton howitzer built by the Krupp factory and used by the Germans to shell Liege and Verdun), military aircraft, poison gas, tank warfare, and all the dreads in the modern nuclear arsenal.

The Industrial Revolution brought to light the perils to society at large of peaceful technological progress. In the wake of rapid advances in agriculture, a surplus population from the farms congregated into vast industrial slums in the big cities. The process that created Manchester and Birmingham and the East End of London was eventually duplicated in the 1940s in Calcutta and Dacca, and is repeated today in the favellas outside Rio de Janeiro and Mexico City.

The polarization between blessing and curse of technology was amplified during the last decades of the nineteenth century. On the one hand there was boundless self-confidence and optimism, a belief in Progress and material culture that seems almost naive to us today. There was also the arrogance and the belief in the superiority of white man that manifested itself in the Prussian superstate and the far-flung colonies of imperial Great Britain. On the other hand there was Marxism, seeing nothing but evil in the capitalist system.

Many modern analysts have expounded the vulnerability of society to technological progress. Perhaps best known is Marshall McLuhan (1911-1980), a Canadian communications theorist who explored the ways television, computers and other media affect society. McLuhan coined the aphorism "The medium is the message," by which he meant that the main impact of new communications technology is not the particular information transmitted via the new media, but how media change society.

McLuhan argued that new technology constitutes "huge collective surgery" carried out on the social body. His favorite example was the motorcar which he blames for the destruction of our cities and the sprawling metropolis. Is it true that the automobile inevitably had to destroy our cities? In theory, it is perfectly feasible to direct highway traffic around a city rather than ploughing through it. You can limit the access of cars to the inner city. A city council faces the task of finding some kind of compromise between on the one hand the pressure of developers and commercial interests that want easy access, and on the other the desire to preserve what living and healthy environment that is still left. Some cities have been more successful than others in finding such a compromise. Many cities are justly proud of their parks, their waterways, and their restoration projects. Not all cities are Los Angeles.

What about the shanty-towns and favellas? Is it true that migration of labor from agriculture to manufacturing and other more labor-intensive occupations inevitably must lead to urban blight? We are here talking about the Western economies during an earlier phase of development when the need for hired labor in agriculture and the wages of such labor was falling, and about the developing world today. The reduced need for hired labor came about and is coming about because of labor-saving technological progress in agriculture. The migration of labor is part and parcel of this technological change. Just as in an individual manufacturing plant progress may occur as one worker is moved from machine A to machine B, progress in the entire economy can occur as one worker moves from one occupation to another. If markets are free, the wage in each occupation will indicate the relative scarcity of workers, and the net return to society of each worker. The falling wages of hired farm workers is a signal telling us that such workers can profitably be moved out of agriculture and into other occupations. The individual worker understands this signal correctly when he leaves his rural occupation and is attracted to urban life. He is attracted by a higher wage. The higher wage signals that the move involves an increase in his net product.

But why then the blight? It is because there is no infrastructure in place to receive the newcomers. A functioning city needs drinking water, sewage systems, electricity, commuter transportation, schools, hospitals, police, a judicial system. To build and maintain these things requires social concensus and social policy. The technological advances thrust the migrants into a social vacuum. The balance between technology and social policy in the original rural setting has been disrupted.

On a deeper level, human balances have been tipped. The move from urban to city life involves something more than just relative wages and net productivity. People loose their roots. The large families living on the farm are broken up. Customs and traditions are lost and individuals are grappling to find a new order in their life, new reference points and a new understanding of their world. Some fail, and the result is shattered dreams, broken marriages, drunkenness, petty crime.

A similar migration of labor from one major category of occupations to another is occurring in the Western world today as the manufacturing sector is shrinking and service occupations attract the surplus labor released from factories. The declining might of the UAW (United Automobile Workers), the large unemployment figures in Detroit, and the falling wages of automobile workers all convey the same message: such workers can profitably be moved out of the American-owned automobile industry and into other occupations. The individual automobile worker understands this signal correctly when he leaves his job in Detroit and looks for an opportunity to be retrained as a copying machine repairman, a hospital waste disposal engineer, or a bank messenger (these are all service occupations). The move will not be easy. A new location, a new residential area, new schools, new friends. It may be a tightrope walk. There are social costs and human costs that are not included in the economic calculation.

Is technology good or bad? Let us look at another example: advances in the medical profession. The great strides that have been taken to conquer infectious diseases, malaria, tuberculosis, polio, and the improvement of health obtained through improved nutrition and preventive medicine would at first sight seem to be indisputably beneficial to mankind. But wait: these victories are upsetting demographic balances in developing nations and ultimately causing the population explosion. In the Western world the falling death figures have gone together with falling numbers of births. The net growth rate of the population has not changed much. But in Asia and Africa and Latin America, the fertility of women has not fallen. The result is a huge positive gap between the birth rate and the death rate.

McLuhan writes: "Any technology gradually creates a totally new human environment. Environments are not passive wrappings but active processes." In the examples that I have discussed, society has turned out to be poorly prepared to handle the advent of the new technology. Engineers and managers thrust a new product or a new way of making things upon a society that possesses no mechanisms for coping and adjusting. Social policy is lagging. But as such policy is being developed, a new equilibrium may eventually be established between technology and society.

It is actually difficult to find examples of new technology that has no adverse effects on society. The only examples that I have been able to think of are technological advances in sports and recreation, and environmental technology.

The stress placed on society can be difficult to trace at first sight. Technological advances in the home -- the dishwasher, the washing machine, microovens and the entire gamut of machinery and gadgets that simplify household chores -- have enabled woman to seek employment outside the home. To many women, this move was "liberation." To others, it spelled misery in a competitive environment that they had not been brought up to handle. A plethora of new social problems presented themselves : finding daycare for preschool children, new pressure in marriages, married couples holding jobs in different cities and spending too little time together, and so on.

How Can Society Manage Technology

The point is not that we do not want technology. It is that we need technology management and social management. Just as technology develops in response to engineering and commercial planning, there is a need to understand at an early stage the need for social change and to develop a social concensus about how such change should be brought about.

The argument is actually quite simple. There exist alternative technological futures. Each such future bends society, for good or bad. We should therefore formulate a technology policy that promotes the development of technologies that shape society in a favorable manner.

There must then also have existed alternative pasts. W. Brian Arthur argues that insignificant chance events often cause one technology to be adopted rather than a competing one. Once the adopted technology has been "locked in," it will be gradually improved through continuous product development. Earlier alternatives fall by the wayside even although they may actually hide potentials for greater long run improvement than the adopted technology. Arthur gives two examples: the steam-versus-gasoline car competition in the 1890s and the nuclear-reactor technology competition of the 1950s and 1960s. In neither case was the outcome obvious. Gasoline was explosive, noisy, hard to obtain in the right grade, and it required complicated new parts. A series of trivial circumstances caused the developers to concentrate their efforts on gasoline. By 1920, gasoline was locked in and steam shut out. Political factors played a role in tipping the scale in favor of the light-water reactor to be used in the U.S.S. *Nautilus*, launched in 1954. Once adopted, learning and eventual construction improvements led the U.S. nuclear industry to favor light water.

Brian Arthur's reasoning agrees with the so-called "butterfly effect" explored in chaos theory. In meteorology, it is the notion that a butterfly stirring the air today by the fluttering of its wings may transform a storm system next month. In scientific terms, it is that a mathematical system may be highly sensitive to its initial conditions. Even small technological decisions may escalate into events that change the world.

This amounts to a real dilemma. Decisions of adopting competing technologies have to be made while we are still in the dark regarding their long run potentials and their long run impact on society. And once a technology has been chosen it becomes locked in place, and the alternatives are never developed. It would have taken prophetic gifts to envision in the 1890s how the automobile were to shape society one hundred years later, and to see the threat to the environment. Nobody could evaluate in 1954 the relative merits of disposing of various kinds of nuclear waste, the danger that radioactive material would come in the hands of terrorists, and other modern day nuclear risks.

Surprises are therefore inevitable. Technologies that appeared to be God-sent blessings may suddenly turn out to involve serious hazards. Only fifteen years ago, most people would have agreed that the advent of central air-condition was a boon to the population that resides in the South. It could be argued that no single technical advance had done as much to make life more pleasant in the temperate zone. That was before the realization that CFCs destroy the ozone layer around the globe.

So, even the most carefully planned technology policy will necessarily have to be complemented by spot repair and patch work, as cracks appear in the technological-societal edifice. Some mistakes can be avoided through careful assessment *ex ante* of technological and social alternatives. Some mistakes have to be corrected *ex post*. Through a process of societal learning -- sometimes bought dearly -- we realize that some technologies have to be curtailed, and others forbidden entirely. In other instances, we shall have to launch "counter-technologies" such as new technology to clean up the environment

In brief, there is a need for government involvment. There is a need for government policy in areas such as occupational safety, public health, the environment, energy, finance, to name just a few. The need for such policy should not be a new theme to the reader of this book. We have encountered it repeatedly, whenever the actions of one individual entrepreneur give rise to social costs ("negative externalities") that are not reflected in his own profit-and-loss statement.

There is a a general issue in this connection that should be addressed right away. It is the matter of *laissez faire* -- the notion that the free market economy is able to deal itself with any of its possible shortcomings, and that the government should keeps its hands off. It is a doctrine that has considerable historical interest. At one time there was political opposition against proposed legislation to prohibit child labor and sweat shops. But once you have accepted the need for *some* government intervention, you have also agreed in principle that the government has a regulatory role to play in the modern economy, laying down the rules of the capitalist game. With a mathematical expression, one might say that the task of enlightened government is to define the "border conditions" inside which capitalism may unfold itself.

I shall discuss government technology policy under five headings: 1. Regulation of technology. 2. Hearings and licencing. 3. "Industrial policy." 4. Putting in place technological "infrastructure." 5. Commercialization and incubation policy.

Regulation

It might be argued that there is little difference between restrictions imposed by nature, and government technology restrictions that really stick. The world of technology is restrictions -- gravity, friction, the second law of thermodynamics; actually, all the "laws" of physics laying down relationships between pressure, temperature, and so on. Physics is strictly ordered and orderly. The art of engineering is to construct a desired object within the confines set by these restrictions. What does it matter if the government adds one or two restrictions of its own? What mathematical economists call "the feasible production set" -- the totality of all doable engineering solutions -- becomes a little bit smaller. A car manufacturer discovers that it is no longer "feasible" to make a line of passenger cars that have a mileage rating of less than 28 miles to the gallon. A pesticide manufacturer learns that a particular compound must not be released into the environment. What does it matter whether it is the government or an engineer who provides management with this information of infeasibility?

The practical problem is to design government regulations that are nondiscriminatory, objective, precise, immediately obtainable and not subject to alternative interpretations. Unfortunately, in the real world, regulation is often the opposite: it is discriminatory, subjective, imprecise, not available until after due administrative process which may take years, and open to various interpretations. Corporate management then faces an entirely new set of costs, both direct costs of collecting information and indirect costs of bureaucratic uncertainty.

I am concerned about the deteriorating quality of new legislation. Roman law was high quality law: precise and succinct. So was Napoleonic law. But the legalese of our day envelops straight-forward matters in a haze of fuzziness. It provides employment for an army of lawyers -- to write it, to interpret it, to challenge it, and to expound on it. And the same fate has befallen government regulation. There exists good regulation and bad regulation. Unfortunately, much of it is poorly designed. Government micro-management of technologies is expensive and should be avoided.

Hearings and Licensing

One way to make regulation easier to enforce and to interpret is licensing: a government agency issues guidelines and technical specifications for a technology; the production and marketing of the technology is licenced as long as it meets the guidelines, but is banned if it does not. In the simplest version of such policy, a blanket licence is issued for a broad class of technologies but an outright ban for others.

One example: In 1987 the U.S. government signed the Montreal Protocol banning chlorofluorcarbons (CFCs) by the end of this century. President Bush later moved up the date to Dec.31, 1995. These chemicals are suspected of destroying the Earth's protective ozone layer. Du Pont Co. has already unveiled a family of refrigerator and air conditioner coolants as substitutes. A consortium of utility companies, advised by the EPA, has launched a contest to build an environmentally friendly refrigerator. Inventors are looking into alternative cooling systems, including a design based on an 1816 invention by the Scottish clergyman Robert Stirling. With the impending ban, a huge multi-billion dollar market of alternatives is beckoning. The multi-splendored engine of creative capitalism rolls into gear. It is perfectly possible that the alternative technologies that eventually will be put in place will involve lower direct costs than the CFC technology. In retrospect, the cost of the regulation may turn out to be a net gain.

The horror example of expensive micro-management is the regulation of the nuclear industry. It has effectively stymied the construction of new nuclear power stations in the U.S. An outright ban on the construction of new nuclear facilities in the U.S. would have achieved precisely the same result. It would have saved an army of inspectors and lawyers, and a mountain of paperwork.

The hearings by the Food and Drug Administration (FDA) and the subsequent approval of new drugs come close to the ideal licencing framework. Consider the epic battle between the two fledgling biotech companies Xoma Corp. and Centocor Inc. They had both developed drugs to treat septic shock or sepsis, a blood poisoning condition. It is the leading cause of death in hospital intensive-care units. The elderly and cancer patients taking chemotherapy are particularly vulnerable. The drug promised to be the first pharmaceutical product available in the U.S. utilizing human monoclonal antibodies -- laboratory-made versions of the antibodies that the body produces to fend off viruses and other foreign substances.

The FDA review process for pharmaceutical products can include a review of a product by an independent advisory committee. In the fall of 1991, such an advisory panel met in Rockville, Maryland to review the competing bids for approval by Xoma and Centocor. The hearing room was jammed with financial analysts, institutional investors and scientists. The analysts were in touch with their home offices via cellular phones. Shares of both stocks were traded in response to the news from the proceedings. During the course of the day, the prices of both stocks gyrated wildly. When the smoke from the gunfight cleared, the panel came down in favor of Centocor, tentatively recommending that its drug be approved, but it put off action on Xoma.

Not a single dose of the drug had yet been sold in the U.S. (Centocor had shipped 200 doses, at $2,500 a dose, to the U.S. Army for use in the Gulf War). But what is traded on Wall Street is not the present. It is a stake in the future. Centocor became the darling of the market. Its price, which only two years earlier had hovered around $10 a share, skyrocketed. At the end of 1991, it stood at $50 a share. The biotech market had become a new Klondyke -- a modern high tech gold rush. The euphoria lasted less than a year. The fate of Centocor well illustrates what happened to the entire industry. New problems with the drug appeared. The FDA requested additional clinical evidence. Eventually, a high number of patient deaths caused Centocor to suspend the clinical trials entirely. The stock price of Centocor plunged. Once again, investors were reminded how fragile the drug-approval process is, and the riskiness of the nascent biotech industry.

There is yet no sepsis drug on the market. Two other biotech companies, Chiron Corp and Synergen Inc also failed to get FDA approval of their experimental drugs. It seems to be the Bermuda Triangle of the biotech industry. But the hearing process works. For a while, a speculative price bubble in a biotech stock can build on Wall Street. But eventually, the speculators must confront the hard facts.

Industrial Policy

How can society manage technology? A third answer is "industrial policy." The term sounds innocuous enough, but what economists and politicians mean by it is pouring government money into industrial projects. The most important examples are the Pentagon and NASA. Defense dollars and space dollars have over the years acted as mighty stimulants of research and development of advanced products in electronics, data processing, guidance systems, robotics, etc.

By definition, industrial policy is not subject to market correction. There are huge risks. Consider the case of U.S. Synthetic Fuel Corp. As the world price of oil exploded in the late 1970s, Jimmy Carter conceived of a plan to create synthetic fuels converting coal into gasoline and a "windfall tax" on oil-company profits to pay for it. Congress approved $20 billion and foresaw the eventual spending of $88 billion, to be doled out to energy companies willing to engage in new energy technology. The Synfuel corporation did finance some projects, but nothing like the original plans. The year Reagan moved into the White House, oil prices began to fall. There is still a strong argument to be made for alternative energy, but few people today would think that synthetic gasoline is it. In 1985, Reagan signed the bill dissolving the Synfuel corporation.

For another example illustrating the pros and cons of industrial policy, consider the race to develop HDTV -- high-definition television. It will have razor-sharp pictures on wide screens. The Japanese have their system in place already, beaming it from a satellite over Tokyo eight hours a day. It was developed by Japan Broadcasting Co., government agencies, and private industry. The Japanese are also developing a flat-panel display screen that is not much thicker than a picture frame. The Europeans have bandied together in a joint venture called Eureka Project 1995. The governments are chipping in 40 % of the cost.

In the U.S., HDTV is still in a planning stage. The domestic TV industry is in ruins. There is just one U.S. manufacturer of television sets left: Zenith Electronics (the sets are assembled in Mexico). Should our government prop up a new generation of electronics companies willing to enter the HDTV arena? Some government funding has been available through Pentagon's Advanced Research Projects Agency (ARPA).

The Federal Communications Commission (FCC) regulates the air waves. It has ruled that any future high-definition TV in this country must be compatible with our present system. This will allow us to watch television on an old set without the improved picture. The FCC has set detailed standards for the future U.S. system. It will be digital. TV shows would be broadcast and received in the form of computer data. A pack of competing electronics companies are preparing for the gradual commercial introduction of HDTV in the U.S. in the late 1990s. The stakes are enormous. Japan Broadcasting Co., pushing the older analog method, has already dropped out of the race. Zenith together with AT&T are taking the global lead in applying the digital language of computers to television broadcasting.

Notice how this success is being achieved: it is done by tapping the ingenuity and creativity of private enterprise *and* by the government setting the technical standards.

Technological Infrastructure

This brings me to the subject matter of technological "infrastructure" , one of the watchwords of the Clinton administration. The Federal highway system, built during the Eisenhower years, was an example of infrastructure put in place by the government that came to have vast beneficial effects throughout the economy, stimulating commerce and industry. In other words, this government investment had massive indirect returns to the economy at large (taking the form of "positive externalities"). Similarly, vice President Al Gore now wants to build an "electronic superhighway." Hopefully, it would form the backbone of a national future communications and education network.

The High-Performance Computing and Communications (HPCC) program is the beginning. It will convert the present Internet -- a chain of 3,000 computer networks linking researchers worldwide -- into a national research and education network. In addition, the proposed $2 billion government spending calls for the construction of new massively parallel supercomputers and the development of the software that would allow them to communicate over long distances. The future vision is a national high-speed network that would reach schools and homes, allowing students to browse electronically through rare books at the Library of Congress while their parents watched television programs with which they could interact.

But could not the very same dream become a reality without government funding, just by setting national standards? Private computer companies and communications companies are already rushing to develop their own systems. In this race, the cable companies are at the forefront. Their immediate goal: five hundred channels of regular television, interactive television, and videos on demand. It will be accomplished through a so-called "set-box," or set-top converter placed on top of a conventional TV set, acting as a gateway between the set and the fiber-optic cable. The box contains the hardware and the software needed by the user to navigate through the system. The box also unscrambles or "decompresses" the arriving pictures -- a technology known as digital compression makes it possible to increase by the tenfold the number of cable channelse that an operator can send to a subscriber's home. Compression standards are being developed by MPEG or the Moving Picture Experts Group, an international body. But the process for building a consensus has been fraught with complications. Each company, or country, has its own political and financial motivations for lobbying the group for various changes.

The largest players in the nascent interactive TV industry are cable giants like Tele-Communications Inc. and Cox Enterprises. Obviously, the financial health of the cable industry will set the pace of its technological advance. In the last instance, that health is determined by the Federal Communications Commission and by the cable rates. Just how crucial those rates are to the momentum of consolidation in the industry was made clear in early 1994, when a roll-back of rates caused two planned major acquisitions to unravel: a proposed $33 billion acquisition of TCI by Bell Atlantic Corp, and a proposed $4.9 billion partnership merging Cox Enterprises and Southwestern Bell.

So, the government really controls all the levers. But, in spite of much lofty talk, there has been little coordinated government policy. Headway in constructing the electronic superhighway will not be made until the government dumps some of its more arcane regulations, including the outdated ban on local telephone companies to provide cable services in their own districts. More competition among cable providers is needed, not less. The barrier between phone services and cable services severely blocks the technical progress. Eventually, the telephone and the TV set will become just two peripherals among many along the superhighway.

In any case, the technological and commercial breakthroughs in this emergent industry are not going to be made by the established players but by startups and new entrants. One of the manufacturers of set-boxes used in current field tests is Scientific Atlanta (SA) Inc., a medium-sized satellite communications company. SA has entered into several strategic alliances that will allow it to help set the standards for new satellite and cable networks, including an alliance with Motorola to provide Earth terminals for the Iridium communications satellite network, an alliance with several Chinese communications companies to provide cable TV to mainland China, and an alliance with cable-giant U.S. West to develop interactive set-top boxes for on-demand video services.

When these lines are written, the interactive industry is still in a state of gestation. To an outside observer, it may look like chaos. There is a bewildering technological diversity, and infinitely many possible commercial combinations. There is technological complexity. And, in this "primal soup," there is evolution. A major evolutionary pace forward is about to be taken. Now is the right moment to formulate wise government policy that can link that evolution into paths that will benefit future generations.

Commercialization Policy

One of the themes of this book has been that the high tech economy requires a rapid flow of commercialization of new technology -- transfers of new technology from the laboratory to the marketplace. We have also seen how commercialization can be facilitated by a series of new organizational setups including high tech incubators, spin-offs, technology alliances, and R&D consortia. While much valuable experience has been gained in this area, the mechanism of technology transfer is still imperfectly understood. Currently, commercialization policy, to the extent that it exists at all in the U.S. today, is still in a search mode.

No doubt this situation is going to change dramatically in the next century, as leading industrial nations embark upon a fullfledged commercialization policy. Such policy will involve government-funded high tech incubators and the agglomeration of many incubators into entire regional techno-cities -- the technopolis model. It will surely accord expanded roles to the Federal research laboratories, charging them not only with basic research but also with advanced product development. It would promote the formation of spin-offs from existing companies, carrying the germs of new technology. In all these applications, the task would be to facilitate and to nourish high tech industry while still preserving one of the basic notions of the free market economy: that all actors in the market, both existing corporations and new entrants, are treated equally and evenly.

Government policy can also be helpful in helping startups to raise funds. During his campaign, President Clinton pledged to explore new ways to securitize bank loans to small businesses into debt securities that can be sold in secondary markets. A proposed agency would buy medium-term small-business loans from banks and finance companies, bundle and repackage them into securities and resell them to institutional investors such as pension funds. And I can see similar government guaranteed credit institutions providing credit to high tech companies -- a kind of high tech development banks. In a fashion, such new credit agencies would parallel the operations that Fannie Mae have already for a long time been carrying out successfully in the mortgage market.

The Swarming of High Technology

The greatest danger of government technology policy is neither the political cost of the government meddling in private businesses nor excessive administrative costs. Rather, it is that government officials might completely miscalculate the current direction of evolution of technology.

The path of technology, leading from the past to the future, is never straight-forward. It cannot be traced by linear extrapolation (although that is precisely what a generation of mathematical economists loved to do, calculating so-called "input-output coefficients" and believing that they stay constant over time). The "current" state of technology always features inconsistencies and contradictions. Instead, with a metaphor brought from the biological world, the path of movement may be likened to that of an insect swarm, such as a swarm of bees. The swarm follows an irregular and zig-zagging path. The swarm as a whole advances forward, but some bees move sideways, or even backwards. An individual bee that at one instant happens to be located at the spearhead of the swarm, may a few seconds later flutter erratically back to the midst of it.

And yet, the movements of the swarm are not random. There is purpose -- the swarm finally descending upon a tree limb located at the other side of the meadow, or even fleeing the beekeeper pursuing the swarm. The swarm is an "intelligent system", displaying a group intelligence that equals more than the sum of intelligence of the individual bees. Using terms of chaos theory that the reader will now be familiar with, the path followed by the swarm is the result of "self-organization."

It is easy to give examples. For a long time in the 1970s and 1980s, industrial robots were thought to be the way of the future. I have earlier in this book described some futuristic robot designs and their market possibilities. And yet, today the business of industrial robots in the U.S. is languishing, with only one remaining manufacturer: Cincinnati Milacron. There is overcapacity in the market for machine tools and computer controls and Cincinnati Milacron is currently phasing out some of its machine tool facilities. It is laying off workers. The market for reprogrammable manufacturing robots is a low-profit area. Instead, the manufacturing industry uses advanced dedicated machinery, like extremely complex spinning machines in the textiles industry, or ever more specialized packing machinery. If one looks at the machine industry as a whole, there is certainly a kind of "group intelligence" directing its path of evolution . There is self-organization and evolution. But the industrial robot is not part of it.

It is the same story with rapid transit trains. In the 1960s the Japanese built their Tokaido Shinkansen system. It is known as the bullet train and is capable of attaining speeds up to 150 miles per hour. It has reduced the travel time between Tokyo and Osaka from six and a half hours to three hours and ten minutes. It carries 124 million passengers a year. In France, a lightweight gas turbine engine was introduced. Known as the "train de grand vitesse" (TGV), it can reach speeds up to 250 miles per hour. It is the fastest operating passenger train in the world. The German electric giant Siemens is developing a similar system for the Bundesbahn (the German railways). With the Eurotunnel (the "Chunnel") under the Channel now being open for traffic, it will soon be possible to travel from London to Rome in a single day, finding time for business conferences in both Paris and Milan along the way. And looking further into the future, an entirely new generation of trains, the "maglev" -- magnetically levitated and propelled trains are already being built in Switzerland. They will experience virtually no friction. Engine, wheels, rails, and rumble will all be eliminated.

And yet, rapid transit is a dead industry in the U.S. There is only one remaining manufacturer of rapid transit trains: Morrison Knudsen, and the company is on the ropes. As the San Francisco Bay Area Rapid Transit District (BART) in early 1995 scaled back its plans to buy additional rail cars, the company ran into a severe cash crunch. It had been plagued by losses in its rail and transit business for years.

These examples, and others like them, tell us an important truth: It is easy to project a series of future life cycles of a technology, one more exciting than the other. But the evolution of technology follows its own inner logic. Yes, there is a swarming of transportation technologies in this country, but those swarms do not include rapid transit. Instead, they include lighter and more fuel-efficient motor vehicles, and short-haul airline operations.

Let me turn to unexpected advances of technology instead, surprise swarmings of novel technology into new and uncharted territory. One would think perhaps that such surprises can no longer occur in a modern and open society where investigative reporters track every step of the top researchers in the nation like bloodhounds, sniffing out their latest ideas and reporting them in the newspaper the next morning. But they do occur, because it takes more than technical reporting to assess the impact of new technology on society.

A good example is word processing. I remember very well reading some eighteen years ago a story in *Time*, explaining the concept at some length. I did not get it. What seems so obvious to every school-child today, was just too radical and novel a concept back then. In those early days nobody could fathom the revolution that word processing would set afoot in every business office in the nation, its impact on daily office routines, on the labor market for clerical staff, and on the role of working women. Back then, the offices of the nation were crammed with underpaid female typists toiling at their electric typewriters. Today, those same workers have been replaced by skilled female specialists doing desktop publishing, spreadsheets and other fancy software. (They are still underpaid, but that is another story.)

Word processing is still exploding in new and unexpected directions. Electronic mail, the Internet, and the World-Wide-Web are but the latest instances of this ongoing evolution.

The next step will be image processing. To achieve the animation and the special effects in his movies, Steven Spielberg uses a battery of computer workstations, splicing together video clips, still pictures, and movie sequences shot from different angles. In the next century my own grandchildren will use those same image blending and splicing techniques in school to do their homework, pasting a video clip downloaded from the Smithsonian directly into their own electronic report books. This amounts to more than just a new learning tool. It opens up a new way of looking at the world around us. People themselves will change.

Even further into the future, I can see a new kind of Internet replacing the current one, providing world-wide interactive image communication. At that point, most business travel will cease. People will live on large estates in the country, commuting electronically to their offices. It will be a very different kind of world.

The Management of the Future

As the twentieth century draws towards its close, many people feel a kind of cultural pessimism -- a feeling that our present civilization has failed to deliver the improved material culture that had been promised us, and that our economy is bogged down in a no-win fight to relieve the damage that was wrought on the environment, damage following from the population explosion, and social disease such as broken homes, drugs, and crime.

Technology does not roll on like a mighty juggernaut, crushing everything under its wheels. Technology is developed by human beings. It develops in response to individual efforts by inventors, designers, business people, and marketing people. What is it that determines the paths followed by the swarming of new technology ? Each corporation and each individual in the complex web of the high tech economy participates in laying down those directions. Technology is what we make it to be.

One of the themes of this book has been the strong ties that exist between technology and the societal setting at large. New technology is born by the ambitions and dreams of men and women; conversely, technology changes society. Ultimately, technology is shaped and formed by the values, norms and priorities of our current civilization. For instance, when people complain about drugs, crime, AIDS, and violence in our society and the huge technology establishments (including illicit operations) that are based on these social ailments, they can all be traced back to the general permissiveness of our civilization. Ultimately, technology policy is a matter of morality.

Technology harnesses the soaring human spirit. Technology built the pyramids in Egypt. It built the Parthenon and the Forum Romanum. It built the great Gothic cathedrals in Europe. It built the pipe organs in those cathedrals. It built castles and palaces of unparalleled beauty. It built the steam engine. It built our entire material civilization. For a long while, the progress of technology was left to the few. Today, it is a mass movement. It does no longer rest on the creativity of a Phidias or a Michelangelo or a Bach. Modern capitalism has found a way of releasing and putting into constructive use the creativity of ordinary men and women. The result is an avalanche of new products and new ways of doing things. For the last few thousand years, the door to the Future used to open ever so little further each year. Suddenly, it is thrown wide open.

Bibliographic Notes

Economic policy is a difficult subject. Contemporary studies of economic policy are typically carried out within the context of some dynamic macro-economic model. My aims here are more limited, my primary interest being focussed on sector policy for which a partial economic model would suffice. But the partial model that I have in mind is complex enough: it needs to spell out the entire logistics network in the manufacturing sector, distribution, and marketing featuring all known technologies and also the expectations about the unknown technologies yet to be developed!

Conventionally, economic policy is formulated in terms of "means" and "goals." The means are parameters like monetary and fiscal policy variables. They could also be environmental controls. The goals might be indicators of the outputs of the industrial sector currently modeled, such as sales, exports, the international competitiveness of corporations, the rate of product development, salaries and wages paid, and so on. They could also be more inclusive concepts like consumer satisfaction, or the quality of life.

Comprehensive as such an analytic format may sound, it would nevertheless fall short of capturing my key argument: that new technologies need to be nurtured, that new organizational forms of collaboration between the private sector and government need to be developed, and that new financial technology needs to be put in place to fund the new ventures.

References

"Catherine the Great brought in an English doctor..." J.T. Alexander, *Catherine the Great*, Oxford University Press, Oxford 1989.

"McLuhan considered that..." M. McLuhan, *Understanding Media: The Extensions of Man*, McGraw Hill, New York 1964.

"McLuhan writes..." *Ibid.*, foreword to paperback edition, p.vi.

"It is the matter of *laissez faire*..." It is true that early neoclassical formulations of interdependent systems assumed that each producer was limited by his "production function" only. But with the advent of modern mathematical equilibrium theory it was realized that the same formulations can be extended to interdependent systems where each producer is limited by any general convex and closed feasible set. See K.J. Arrow and G. Debreu, "Existence of an Equilibrium for a Competitive Economy," *Econometrica*, July 1954, pp. 265-290. For a general discussion of *laissez faire*, see R. Kuttner, *The End of Laissez-Faire: National Purpose and The Global Economy after the Cold War*, Alfred A. Knopf, New York 1991

"Inventors are looking into alternative cooling systems, including a design based on an 1816 invention " The company is Carrier Corp. owned by United Technologies Corp.

"Consider the epic battle between the two fledgling biotech companies..." Most of information concerning Centocor and Xoma I have brought from a prospectus for Tocor II, a new research arm of Centocor underwritten by PaineWebber Incorporated, The First Boston Corporation, Hambrecht & Quist, and J.P.Morgan Securities Inc.

"For another example illustrating the pros and the cons..." C. Carbonara, "HDTV," in *The United States and Japan: Shared Progress in Technology Management*, ed. by S. El-Badry, H. Lopez-Cepero, and F. Y. Phillips, IC^2 Institute, University of Texas at Austin, 1993.

"W. Brian Arthur argues..." W. Brian Arthur, "Competing Technologies, Increasing Returns, and Lock-in by Historical Events," *The Economic Journal*, March 1989.

"Brian Arthur's reasoning agrees with the so-called 'butterfly effect'..." See J. Gleick, *Chaos: Making a New Science*, Penguin Books, New York 1987.

"...the path of movement may be likened to that of an insect swarm." I have read Mark M. Millonas, "Swarms, Phase Transitions, and Collective Intelligence," *Proceedings of the Artificial Life III Conference* held June 15-19 in Santa Fe, Santa Fe Institute, New Mexico.

"And looking futher into the future, there are the possibilities of an entirely new generation of trains..." J. Neffe, "German Maglev Trains Way Ahead of Rivals," *Nature*, Sept. 1988.

References

IC^2 Publications

Reference [1]: G. Kozmetsky, M.D. Gill, Jr. and R.W. Smilor, *Financing and Managing Fast-Growth Companies: The Venture Capital Process*, Lexington Books, Lexington, Mass. 1985.

Reference [2]: G. Kozmetsky, *Transformational Management*, Ballinger Publishing Co., Cambridge, Mass. 1985.

Reference [3]: R.W. Smilor and M.D.Gill, Jr. (editors), *The New Business Incubator*, Lexington Books, Lexington, Mass. 1986.

Reference [4]: J.R.Kirkland and J.H. Poore (editors), Supercomputers, Praeger, Westport, Conn. 1987.

Reference [5]: T.J. Mabry (editor), *Plant Biotechnology: Research Bottlenecks for Commercialization and Beyond*, IC^2 Institute, The University of Texas at Austin 1987.

Reference [6]: S. Nozette and R.L. Kuhn (editors), *Commercializing SDI Technologies*, Praeger, Westport, Conn. 1987.

Reference [7]: M.J. Petit, *Industrial Research & Development Consortia*, IC^2 Institute, The University of Texas at Austin 1987.

Reference [8]: A. Furino (editor), *Cooperation and Competition in the Global Economy*, Ballinger Publishing Co., Cambridge, Mass. 1988.

Reference [9]: G.R. Bopp (editor), *Federal Lab Technology Transfer*, Praeger, Westport, Conn. 1988.

Reference [10]: R.W. Smilor, G. Kozmetsky and D.V. Gibson (editors), *Creating the Technopolis*, Ballinger Publishing Co., Cambridge, Mass. 1988.

Reference [11]: R.L. Kuhn (editor), *Frontiers of Medical Information Sciences*, Praeger, Westport, Conn. 1988.

Reference [12]: K.D. Walters (editor), *Entrepreneurial Management: New Technology and New Market Development*, Ballinger Publishing Co., Cambridge, Mass. 1989.

Reference [13]: T.J. Mabry, S.C. Price and M.D. Dibner, Commercializing *Biotechnology in the Global Economy*, IC^2 Institute, The University of Texas at Austin 1991.

Reference [14]: F. Phillips, editor, *Concurrent Life Cycles*, IC^2 Institute, The University of Texas at Austin, Texas 1992.

IC^2 Publications, Series on Econometrics and Management Science

Reference [15]: V. Mahajan and Y. Wind (editors), *Innovation Diffusion Models of New Product Acceptance*, Ballinger Publishing Co., Cambridge, Mass. 1986

IC² Publications, International Symposia in Economic Theory and Econometrics

Reference [16]: W.A. Barnett, J. Geweke, and K. Shell (editors), *Economic Complexity: Chaos, Sunspots, Bubbles, and Nonlinearity*, Cambridge University Press, Cambridge, Mass. 1989

IC² Publications, Quorum Books

Reference [17]: R.A. Peterson, G. Albaum, and G. Kozmetsky, *Modern American Capitalism: Understanding Public Attitudes and Perceptions*, Quorum Books, Westport, Connecticut, 1990.

Reference [18]: S. Thore, *Economic Logistics: The Optimization of Spatial and Sectoral Resource, Production, and Distribution Systems*, Quorum Books, Westport, Connecticut, 1991.

Reference [19]: Y. Ijiri (editor), *Creative and Innovative Approaches to the Science of Management*, Quorum Books, Westport, Connecticut, 1993.

Reference [20]: R.L. Kuhn (editor), *Generating Creativity and Innovation in Large Bureaucracies*, Quorum Books, Westport, Connecticut, 1993.

Index of Corporations and Executive Officers

3M, 10, 18

Advanced Micro Devices, 48
Akers, John, 4
Allen, Robert, 30, 79
Allied Stores, 81
Amdahl, 4, 32, 55
American Cyanamid, 80
American Diversified, 168, 177
American Home Products, 80
Ameritech, 30
Amgen, 120
Ampex, 33
Apollo, 4, 13, 151
Apple Computer, 34, 54, 55, 144, 147, 151, 153
Asea BrownBovery, 77
Astra AB, 121
AT&T, 17, 29, 30, 42, 55, 63, 78, 79, 80, 151, 154, 187
Atari, 54, 144

Bank of New England, 174
BankAmerica, 174
Bankers Trust, 162
Beatrice, 80
Bell Atlantic, 30, 189
Bethlehem Steel, 3
Biogen, 120
Blockbuster Entertainment, 80, 91
Borland, 4
British Petroleum, 80

C&S Sovran, 174
Cable News Network, 71, 73
Calgene, 117
Campeau, 80, 81

Carlson, Chester, 29, 98
Carnegie, Andrew, 2
CBS, 64, 72
Centel, 30
Centocor, 185, 186, 194
Cetus, 118, 120, 124
Chaparral Steel, 3
Chemical Banking, 174
Chiron, 120, 124, 186
Cincinnati Milacron, 191
Citicorp, 56, 157
Columbia Pictures, 80, 81
Compaq Computer, 54, 55
Conner Peripherals, 54
Contel, 80
Control Data, 145, 146
Cormetech, 22
Corning, 21
Cox Enterprises, 189
Cray Computer, 32
Cray Research, 32, 161
Cyrix, 10

Data General, 55
Dell Computer, 54
Digital Equipment, 55, 146
Disney, 37, 46, 58, 77
Dow Jones, 32, 108
Drexel Burnham Lambert, 155, 158
Du Pont, 118, 185

Eastern Airlines, 90
Eastman Kodak, 29, 78, 80
Electrolux, 77
Eli Lilly, 118, 119
ETA Systems, 32
Exley, Charles, 79
Exxon, 79, 129, 137

Fairchild Semiconductor, 143
Farmers' Group, 80
Federal Express, 68
Federated Department, 81
First Boston, 170, 194
Fisher, George, 78
Floating Point Systems, 54
Ford Motor, 151
Ford, Henry, 2
Frick, H.C., 2
Fujitsu, 20, 32

Genentech, 1, 119, 120, 121, 123
General Dynamics, 11, 40
General Electric, 80
General Foods, 80
General Motors, xi, 4, 77, 80
Gerstner, Louis, 4
Getty, Paul, 71
Goldman Sachs, 32
GTE, 80
Gulf+Western, 83

Haloid, 98
Harris, 146
Hewlett, William, 144
Hitachi, 32
Honeywell, 146
Hughes Aircraft, 23, 80

IBM, xi, 4, 10, 17, 20, 21, 39, 55, 146, 147, 148, 149
IG Labs, xi
Intel Corp, 1, 48, 148
Interactive Systems, 78
ITT, 83

Japan Broadcasting, 187
Jobs, Steven, 144, 151
Johnson, Ross, 87, 88

Kabi Farmacia AB, 121
Kohlberg Kravis Roberts, 80, 87, 88

Kozmetsky, George, 23, 24, 25
Kraft, 80
Kravis, Henry, 88

Litton Industries, 23, 82
Lotus, 4

Macy's, 81
Manufacturers Hanover, 174
Marathon Oil, 3
Maxxam, 90
McCaw Cellular, 79, 161, 162
McDonnell Douglas, 20, 40
McGowan, William, 161
MCI, 30, 161
Medco Containment, 80
Merck, 80
Metro Mobile, 30
Metromedia, 30
Microsoft, 1, 4, 9, 20, 34, 42, 55
Milken, Michael, 158, 159, 161, 163
MIPS, 10, 31
Mitsubishi, 21
Mitsui, 21
Mobil Oil, 80
Mobile Communications, 30
Montgomery Ward, 64
Morrison Knudsen, 191
Motorola, 20, 78, 144, 145, 149, 189

Nabisco, 4, 80, 87, 88, 92, 162
National Cash Register, 79
National Computer, 54
National Semiconductor, 145
NBC, 72
NCNB, 174
Nestle, 77
Nomura Securities, 32
Norris, W.C., 146
Noyce, Robert, 143
Nucor Corp, 3
Nynex, 30

Pacific Telesis, 30

Index

Packard, David, 144
Paramount Communications, 75, 82, 91
Pennsylvania Railroad, 46
Philips, 29
Polaroid, 29

Quantum, 54
QVC Network, 76, 91

RCA, 27, 33, 42, 72, 80, 146
Reed, John, 56
RJ Reynolds, 81, 87
RJR Nabisco, 4, 80, 87, 88, 92, 162
Ross, Steve, 82
Royal Dutch Shell, 80

Safeway Stores, 80
Salomon Brothers, 170
Scientific Atlanta, 189
Seagate Technology, 54
Sears and Roebuck, xi
Security Pacific, 174
Sharp, 20
Shearson Lehman Hutton, 88
Shell Oil, 80, 126, 140
Siemens, 148, 191
Silicon Graphics, 10, 54, 161
Singleton, Henry, 23
Sony, 27, 29, 58, 80, 81
Southland, 80
Southwestern Bell, 30, 189
Sperry, 146
Standard Brands, 87
Standard Oil, 76, 80
Sterling Drug, 80, 81

Stores, 80, 81
Sumitomo, 21
Sun Microsystems, 10, 55, 147, 151, 154
Symbol Technologies, xi
Systemix, 119

Tandon, 54
Teledyne, 23
Texaco, 71
Thinking Machines, 32
Thornton, C.B., 23
Time, 50, 80, 81, 82, 96, 101, 192
Time Warner, 81
Tseng Labs, 10
Turner Broadcasting, 161
Turner, Ted, 71, 72

U.S. Sprint, 29
U.S. Steel, 2, 3
U.S. West, 30, 189
Unisys, 55
USX Corporation, 3

Viacom, 76, 80, 91

Warner, 80, 81, 82
Wedtech, 152, 154
Western Union, 8, 171
Westinghouse, 72

Xerox, 98
Xoma, 185, 186, 194

Zenith Electronics, 187

About the Author

Sten Thore is the Gregory A. Kozmetsky Centennial Fellow in the IC^2 Institute, the University of Texas at Austin. He is a senior research scientist at the university. He teaches in the departments of economics ("Economic change and creativity") and aerospace engineering ("The commercialization of space technology"). He is a faculty member of the new Executive M. Sc. program in the Commercialization of Science and Technology that the IC^2 Institute will be teaching in Washington D.C., starting in the fall of 1995.

Since Dr. Thore joined the IC^2 Institute in 1978, he has been working on resource and supply systems modeling, industry logistics and the economics of high technology. He has recently published two books: a monograph entitled *Economic Logistics: The Optimization of Spatial and Sectorial Resource, Production, and Distribution Systems* (Quorum Books, Westport, Conn., 1991), and a textbook entitled *Computational Economics: Economic Modeling with Optimization Software* (The Scientific Press, South San Francisco, Calif. 1991, coauthored with G.L. Thompson of Carnegie Mellon University).

Dr. Thore has authored or coauthored nine books and more than 80 research papers. His recent contributions include the development of a new constrained least squares regression technique (with applications to productivity change in the manufacturing sector), stochastic formulations of so-called data envelopment analysis (with applications involving the comparison of the efficiency of capitalism and state socialism), the pricing of heterogeneous goods (such as high technology products with many consumer attributes), and several studies of the cost effectiveness and competitiveness of the U.S. computer industry.

Before coming to Texas, Dr. Thore held a chair in economics at the Norwegian School of Economics and Business Administration, Bergen, Norway. At this time he specialized in the optimization of bank funds management, financial intermediation, and the flow-of-funds. On various sabbatical leaves, he was a visiting professor to Northwestern University, Carnegie Mellon University, and the University of Virginia.

Dr. Thore holds the degree of *filosofie doktor* from the University of Stockholm, Sweden. He was a founding member and the first chairman of the Norway Chapter of The Institute of Management Sciences. He is a naturalized U.S. citizen; he was also commissioned an honorary citizen of the state of Texas. He is listed in *Who is Who in the World*.

Address: IC^2 Institute, 2815 San Gabriel, Austin, Texas 78734, telephone 512-478-4081.